To Gerhard Haxel mit ganz vielem Dank für die tolle mit Unterstützung und Aloha aus Hawaii,
Dieter Mueller-Dombois
Juli 2013 Kailua, HI

ʻŌhiʻa Lehua Rainforest

Für Prof. Gerhard Haxel
mit herzlichstem Dank für
all Ihren Einsatz für meine
Karriere und allen guten
Wünschen!
In tiefer Verbundenheit
Ihr
[illegible]

D1641674

Library of Congress Cataloging-in-Publication Data

Mueller-Dombois, Dieter, 1925–
'Ōhi'a Lehua rainforest: Born among Hawaiian volcanoes, evolved in isolation; The story of a dynamic ecosystem with relevance to forests worldwide / Dieter Mueller-Dombois, James D. Jacobi, Hans Juergen Boehmer, Jonathan P. Price.
p. cm.

Photos by Dieter Mueller-Dombois, unless otherwise noted
Cover design by Marween Yagin
Book design by Mark Nakamura

ISBN-13: 978-0615744353 (Custom Universal)
ISBN-10: 0615744354

© 2013 Copyright Friends of the Joseph Rock Herbarium

All Rights Reserved. No part of this book may be reproduced in any form or by any electronic or mechanical means, including information storage and retrieval devices or systems without prior written permission from the publisher, except that brief passages may be quoted for reviews.

ʻŌhiʻa Lehua Rainforest

Born Among Hawaiian Volcanoes, Evolved in Isolation

The Story of a Dynamic Ecosystem with Relevance to Forests Worldwide

Dieter Mueller-Dombois
University of Hawaiʻi at Mānoa

James D. Jacobi
U.S. Geological Survey

Hans Juergen Boehmer
Technical University of Munich

Jonathan P. Price
University of Hawaiʻi at Hilo

Preface by
Samuel M. ʻOhukaniʻōhiʻa Gon III

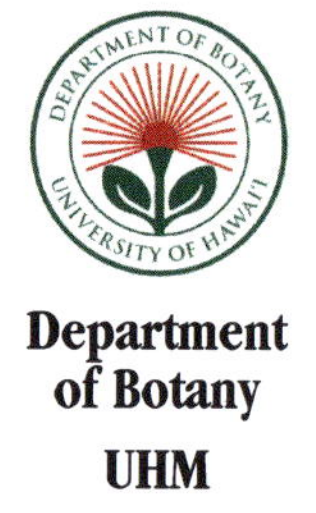

Deutsche
Forschungsgemeinschaft

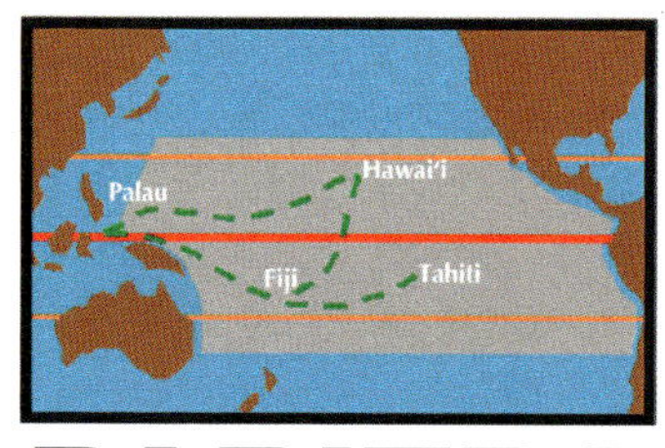

Contents

Dedication

We dedicate this book to five outstanding field researchers and former graduate students in the University of Hawai'i at Mānoa (UHM) Department of Botany who since passed away, but whose contributions played a significant role in the study of the Hawaiian 'Ōhi'a rainforest ecosystem as here presented. They include:

Garrett A. Smathers (PhD 1972)

Former Park Naturalist, who after earning his PhD advanced to become Chief Scientist of the US National Park Service. In that position he helped secure our first major research grant in 1977. As an alternative to the disease hypothesis for explaining stand-level dieback, that first major grant was awarded for the competing idea that canopy dieback could be a normal process in succession, an idea that other grant authorities thought to be unpromising.

Ranjit G. Cooray (MSc 1974)

Who worked as a research assistant in the Hawai'i IBP and measured numerous stands of 'Ōhi'a trees in dieback and non-dieback condition that led to the discovery of their nature as cohort stands. He also drew the forest stand profiles shown as Figures 1.1 and 1.2 in Chapter 1 of this book.

Nengah Wirawan (PhD 1978)

His dissertation research of a tropical rainforest on windward Oʻahu provided for the ecological baseline of a PABITRA site, a summit to sea landscape of historic significance, the Kahana Valley Ahupuaʻa (*Pacific Science* 59(2): 283–314). Nengah's contribution, which in addition to his focus on soil-water regimes, includes a vegetation map and a plant checklist, is the last paper (posthumously published) in an entire issue devoted to the PABITRA network. This Kahana Valley Ahupuaʻa, which became a State Park after Nengah's first geobotanical involvement, now represents Hawaiʻi on the PABITRA logo (see p. iv).

Nadarajah Balakrishnan (PhD 1985)

He took great interest in the study of soil relationships in dieback and non-dieback stands and analyzed numerous soil samples for pH, organic matter and nutrients in relation to our habitat classification (*Pacific Science* 37: 339–359).

Lani Stemmermann (PhD 1986)

She discovered that ʻŌhiʻa tree populations change in morphology and physiology along the primary successional gradient (*Pacific Science* 37: 361–373). She co-authored (with John Dawson, NZ) the taxonomic revision of the Hawaiian *Metrosideros* species and variety complex, and established a Common Garden for the study of altitudinal and successional ecotypes. That Common Garden still functions as a resource for genetic research. It spawned several research papers after Lani's untimely death, and it remains as one of her many legacies.

Acknowledgements

In addition to the graduate students in vegetation/ecosystem ecology who completed their thesis research on the ʻŌhiʻa lehua rainforest (Appendix A), we feel obligated to acknowledge all those colleagues who contributed research time and thought effort on the ʻŌhiʻa forest dieback/decline problem.

They include among Hawaiʻi State Foresters Edwin Q. P. Petteys (who did the first aerial photo survey of ʻŌhiʻa rainforest decline), Tom Tagawa, and Libert Landgraf (concerned successive Hawaiʻi State Foresters); among US Forest Service Research Foresters Robert E. Nelson (Head of USFS Institute of Pacific Islands Forestry, who secured funding for the disease hypothesis and arranged for an aerial photo survey that resulted in a colored air photo set at the scale of 1:12,000 covering 100,000 ha), Robert E. Burgan, F. F. Laemlen, and Robert V. Bega; among pathologists Ian Buddenhagen, Oliver V. Holtzmann (successive Chairs of the UHM Plant Pathology Department), Charles S. Hodges (mycologist and Director of USFS Institute of Pacific Islands Forestry), Robert F. Scharpf, Richard S. Smith, Jr., John T. Kliejunas, Wen-Hsuing Ko, S. C. Hwang (the latter three became involved in intensified pathological research clarifying that the root pathogen *Phytophthora cinnamomi* was not the killer fungus as earlier suspected), F. J. Newhook (called in as adviser from New Zealand), and Martin F. Stoner (mycologist and member of the Island IBP research team); among field entomologists Richard P. Papp, John D. Stein (both USFS entomologists), G. A. Samuelson, J. Linsley Gressitt (both Bishop

Museum entomologists and members of the Island IBP research team, and Jack W. Beardsley (insect pest specialist from UHM entomology group and member of the Island IBP team); among US forest ecologists and soil scientists Kenneth T. Adee, Hulton Wood (both did forest stand measurements and soils research and classified the rainforest dieback into essentially the same substrate and structural types as the UHM Botany team), Robert D. Doty (USFS hydrologist who monitored soil water fluctuations, stream flow, and did a climate analysis), C. John Ralph (USFS ornithologist), B. R. McConnel, and Paul Higashino (field assistant to Ken Adee and Hutch Wood).

The UH Mānoa Botany graduate students directly involved in the dieback research included in addition to Jim Jacobi (co-author of this book), Ranjit Cooray, Nadarajah Balakrishnan, Richard Becker, Grant Gerrish, Phil Burton, Lani Stemmermann, Joan Canfield, Wayne Takeuchi, Alan Holt, Kanehiro Kitayama, Mallikarjuna Aradhya, Donald Drake, and Yoshiko Akashi. Substantial field research support came from visiting faculty including William E. Evenson (Brigham Young University) in1982/83 and from Reinhard F. Huettl (University of Freiburg, Germany and Head of Research, Kali and Salz Kassel) in 1990/91.

Among politicians who supported us in getting our first proposal for funding approved, we acknowledge Hawai'i County Council member Merle K. Lai, the honorable US Senator Hiram L. Fong, and the Hawai'i State Chair of the Ecology, Environment & Recreation Committee, the honorable Jean King (1975–1977).

We could not have done our research without financial support. We received two major grants, the first from the National Park Service (Contract CX 8000-6-006 $55,000 April 1977), the second from the National Science Foundation (NSF DEB-7910993 $425,000 1979–1983). We also received supplementary grants from McIntire-Stennis Haw-00684 $15,000 1982–1985 and a follow-up grant of $25,000 1985–1987; East-West Center-UH Collaborative Research Grant $14,000, 1983–1984. Subsequently, a starting grant of $3000 was received from the Tuexen Foundation in the year 1999 for research travel to initiate postdoctoral research for Hans Juergen Boehmer, who became the second co-author of this book. His postdoctoral research 2001–2003 was funded by a grant from the German Research Foundation (DFG) (DFG BO-1768-1 $85,000), and another grant from the Arbeitsgemeinschaft Freiraum, Kassel, Germany ($10,000). Support for J. Jacobi was provided by the US Geological Survey's Pacific Island Ecosystems Research Center under the Ecosystems Program. Any use of trade, product, or firm names is for descriptive purposes only and does not imply endorsement by the US Government.

Acknowledgement is due also to our peer reviewers Kent W. Bridges, Curtis Daehler, Kasey Barton, László Orlóci, Peter Vitousek, Michael Barbour, Linda Pratt, Joan Canfield and Linda Mertelmeyer, who took the time to read through various drafts of our manuscript. We thank them for their constructive criticism, and their remarks helped substantially in completing our final version of the book.

Mark Nakamura and Marween Yagin (UHM Center for Instructional Support) guided us with their technological and artistic expertise throughout the production process of this book. They designed the book cover and book flyer and made several other useful suggestions for the layout of this book. Their kind personality and professional dedication contributed substantially to the timely completion of our book project. We also thank Samuel M. ʻOhukaniʻōhiʻa Gon III for writing the preface chapter.

For all this support and collaboration, which also includes the sustained and most valuable assistance of Annette Mueller-

Dombois, the first author's wife, we are sincerely grateful. Without it we could not have overcome the many challenges that led to completing this book.

Dieter Mueller-Dombois
Kailua, Hawai'i
September 12, 2012

Preface

'Ōhi'a Lehua (*Metrosideros polymorpha*), the dominant tree of the Hawaiian rainforest, carries immense cultural significance. The tree is considered one of the *kinolau*, a physical manifestation of Kū, one of the four principal Hawaiian deities.

Kū stands for strength and anchor, something to hang onto. This can be translated into the ecological concept of keystone species, a dominant species that holds an ecosystem together.

Lehua, the red, orange and yellow flowers, serve as the fiery symbol of Pele, goddess of fire. The first line to a well-known chant to Laka, deity of the hula, says:

Noho ana Laka i ka uluwehiwehi,
Laka resides in thick forest growth,

ka 'Ōhi'a kū i luna o Mauna Loa.
the tall 'Ōhi'a of Mauna Loa.

Furthermore, the foliage and blossoms of lehua are appropriate offerings on the *kuahu hālau*, the shrine for the traditional schools of hula.

Lehua, the blossom of the 'ōhi'a tree, carries many significant meanings. The crimson red of lehua symbolizes the blood of warfare. The first warrior to fall in battle is called Lehua. Thus 'Ōhi'a lehua symbolizes strength in life and in death.

The wood of 'Ōhi'a lehua is incorporated into some of the most sacred structures of the *heiau* (temples) of governance:

the *lele* (offering platform), the *'ānu'u* (oracle tower), and the *ki'i akua* (god figures). The temple images of Kū, held such high mana (spiritual power) that it required a human sacrifice to remove an 'Ōhi'a from the forest if it was destined to become the likeness of Kū. Even the *pā* (enclosures) that surrounded temple sites were fashioned out of 'Ōhi'a boles. For its superior strength, the wood of 'Ōhi'a was preferred for weapons such as *lā'au* (war clubs) and *pāhoa* (daggers). The same properties applied to tools such as *'ō'ō* (agricultural digging sticks) and *ko'i* (adzes).

In traditional medicine (*lā'au lapa'au*), the tender *liko lau*, or leaf buds, and the *a'a lewa*, or aerial roots of 'Ōhi'a, were used in a variety of ways, as a tonic and intervention for failed appetite, and as treatment for various infections.

As a prevalent and easily recognized presence in ecosystems from sea level to tree-line, numerous *'ōlelo no'eau* (wise and poetical sayings) refer to 'Ōhi'a as a positive symbol of strength, sanctity, and beauty. *'Ōlelo no'eau* are often derived from *oli* (chant) and *mo'olelo* (story), and the frequency of mention in such *mo'olelo* is an indication of significance. In the well-known epic of Hi'iaka-i-ka-poli-o-pele, the 'Ōhi'a lehua is the single most frequently mentioned plant, far outstripping any other plant save hala (*Pandanus tectorius*).

In the *Hawaiian Dictionary* is mentioned the widely-known and oft-quoted warning that picking lehua blossoms will bring rain. The tree is the absolute dominant in the wet forest ecosystem. Thus, it is not surprising that the tree and the flower are both strongly associated with rain. Different kinds of rain, recognized by the early Hawaiians, refer to lehua, such as the *Ua-kani-lehua* (rain that quenches the thirst of the lehua), the *Moaniani-lehua* (wafting lehua fragrance), and the *Līlī-lehua* (cold lehua drizzle).

Wetter districts, such as Hilo and Puna on the Island of Hawai'i, strongly identify with the 'Ōhi'a lehua, and the wet 'Ōhi'a forests of Kohala have been called *lā'au 'ohi wai*, the forest that gathers water, linking in the minds of Hawaiians that the name ʻŌhiʻa alludes to the gathering (*ʻohi*) of the water of life, high in the realm of the gods.

All of these primary cultural underpinnings make ʻŌhiʻa lehua perhaps the single most culturally significant tree in Hawaiʻi. This weighs heavily when considering the value of ʻŌhiʻa as an element of essential conservation value in Hawaiʻi. Let this foundation linger with you as you explore the study of the dynamics of ʻŌhiʻa lehua forest: born among volcanoes, evolved in isolation and carrying a message with relevance to forests worldwide.

Samuel M. ʻOhukaniʻōhiʻa Gon III, PhD
Senior Scientist/Cultural Advisor
The Nature Conservancy of Hawaiʻi

Photo courtesy National Park Service

Chapter 1

A Mature Hawaiian Rainforest

A first look at a mature ‘Ōhi‘a/tree fern forest

The mature Hawaiian rainforest presents itself in its natural beauty and tranquility when entering Hawai‘i Volcanoes National Park (Photo 1.1). You can follow the rainforest on both sides of the Crater Rim Road (Photo 1.2, 1.3, and 1.4). Earlier signs along this road referred to this forest as

Photo 1.1. The Hawai‘i Volcanoes National Park entrance sign on Highway 11. Here you can see native Hawaiian rainforest on both sides of the road.

Photo 1.2. Mature 'Ōhi'a lehua (*Metrosideros polymorpha*) rainforest. Note the uniform structure of the trees. They represent a cohort stand, a generation of trees that became established after a major eruption of Kīlauea volcano in 1790.

Photo 1.3. Another stand segment near Park entrance. The bright green bush in center is a young 'Ōlapa (*Cheirodendron trigynum*), a common subcanopy tree in the 'Ōhi'a rainforest.

Fern Forest, on account of its striking undergrowth of Hāpuʻu tree ferns (*Cibotium* spp.). But the overstory is dominated by only one tree species, the ʻŌhiʻa lehua (*Metrosideros polymorpha*) (Photo 1.5). Both the tree ferns and the ʻŌhiʻa lehua are endemic species.

Photo 1.4. Mature ʻŌhiʻa lehua/Hāpuʻu tree fern forest along Crater Rim Road in Hawaiʻi Volcanoes National Park.

Photo 1.5. Flowers of the ʻŌhiʻa lehua (*Metrosideros polymorpha*), the dominant tree species in Hawaiian rainforests.

The word "endemic" means these species occur nowhere else in the world—in this case only in the Hawaiian Islands. Here they evolved as new species from ancestors that arrived naturally through long-distance dispersal. The small spores of the tree ferns most likely came with the jet stream from tropical America. The light wind-born seeds of 'Ōhi'a trees probably came with rare storm events from the Society Islands or the Marquesas, where its closest relative *Metrosideros collina* is at home. Other propagules, larger than from ferns and the 'Ōhi'a tree, came in the guts of birds or attached to their feathers. That includes all native species with fruits. A small number of plant species found near beaches in the coastal areas arrived on sea currents. Such long-distance events that establish ancestor species in Hawai'i are extremely rare, but they can still occur. If natural colonizing events from overseas would have been more frequent, the colonizing ancestors would not have evolved into new endemic species.

There are other native species, such as the Uluhe fern (*Dicranopteris linearis*), which can also be seen along the Crater Rim Road, usually where the tree canopy is more open (Photo 1.6). This fern forms thick mats and occasionally climbs halfway up the 'Ōhi'a tree trunks. This Uluhe fern is also found in other Pacific islands, for example in Fiji. Such species, which have not changed significantly from their ancestors growing outside Hawai'i, are called indigenous. They are not unique to Hawai'i.

COLONIZATION BY 'ŌHI'A

Recent research using molecular genetics methods puts the arrival of 'Ōhi'a lehua as approximately 5 million years ago (Percy et al. 2008). This ancient date coincides roughly with the volcanic emergence of Kaua'i, the oldest high island in the Hawaiian chain. The arrival time of the Hāpu'u tree fern is not yet known, but it may be a similarly ancient member in the Hawaiian flora.

Photo 1.6. Open patch in ʻŌhiʻa rainforest with mat-forming Uluhe fern (*Dicranopteris linearis*), a heliophytic (light loving) fern. Under closed tree cover the shade-tolerant Hāpuʻu tree ferns (*Cibotium* spp.) prevail as shown on Photos 1.2 and 1.4.

Photo 1.7. ʻŌhelo kau lāʻau (*Vaccinium calycinum*), another common endemic shrub in the ʻŌhiʻa rainforest understory.

But since they became established in Hawai'i naturally, that is unaided by humans, they also are native species.

Looking closer into the Hawaiian rainforest one will notice other plant species, for example, lower growing ferns, such as the just mentioned Uluhe fern or the 'Ōkupukupu fern (*Nephrolepis exaltata*), both native species. Together with the tree ferns you find some native shrubs, such as the 'Ōhelo kau lā'au (*Vaccinium calycinum*) (Photo 1.7) and 'Ilihia (*Cyrtandra platyphylla*) (Photo 1.8), and also some native trees, such as Pilo (*Coprosma ochracea* and *C. rhynchocarpa*), Kāwa'u (*Ilex anomala*) (Photo 1.9), and 'Ōhā wai nui (*Clermontia hawaiiensis*) (Photo

CYRTANDRA is the richest genus of the African violet family (Gesneriaceae) with 60 endemic species in the Hawaiian islands. Most of them, like the 'Ilihia in Photo 1.8, are shrubs growing in understory of the native rainforest. Members of this genus are found throughout the Pacific islands.

Photo 1.8. A flowering 'Ilihia (*Cyrtandra platyphylla*), one of the more common endemic shrubs in the undergrowth of this rainforest.

1.10). In any given area of Hawaiian rainforest about the size of a houselot, one can expect to find about twenty to thirty additional native plant species, many of which are smaller ferns, and more can be seen where human-introduced species have invaded.

The ʻŌhiʻa forest also provides important habitat for a unique assemblage of native Hawaiian birds, all of which are found nowhere else in the world but Hawaiʻi (Photo 1.11). Many of these species are closely related, belonging to the en-

Photo 1.9. A young Kāwaʻu (*Ilex anomala*), a common native subcanopy tree in the Ōhiʻa rainforest. The Kāwaʻu is in the center growing above the Hapuʻu tree fern.

THE BELLFLOWER FAMILY is one of the most spectacular groups in the Hawaiian flora in terms of endemism. It contains five endemic genera (*Brighamia, Clermontia, Cyanea, Delissia, and Trematolobelia*). *Cyanea* is the most diverse genus with 60 species, *Clermontia* is next with 22 species, most of them undergrowth shrubs in the Hawaiian rainforest such as the one shown in Photo 1.10. *Brighamia* is a rare shrub with just two species that grow on wind-exposed cliffs on Kauaʻi and Molokaʻi.

Photo 1.10 (left). 'Ōhā wai nui (*Clermontia hawaiiensis*) one of the more common native species in the Bellflower family (Campanulaceae).

Photo 1.11. Some of the native forest birds known from the main Hawaiian Islands: (1) Palila, (2) 'Apapane, (3) 'I'iwi, (4) 'Amakihi, (5) 'Ōma'o, (6) 'Elepaio, (7) Po'o-uli, (8) 'Akiapōla'au, and (9) O'o'a'a. Only the 'Apapane, 'I'iwi, 'Ōma'o, and the 'Amakihi are still commonly seen in the National Park. The Palila is strictly confined to the high-elevation forest on Mauna Kea and the other five birds are now very rare. The 'O'o'a'a and Po'o-uli were last seen as recently in the 1980s but both are now presumed to be extinct. Painting copyright ©1975 by H. Douglas Pratt, reprinted with permission.

demic Hawaiian Honeycreeper family. These include two nectar feeders, the ʻApapane (*Himatione sanguinea*), the most abundant bird in all Hawaiian forests, and the red ʻIʻiwi (*Vestiaria coccinea*) with its crimson colored feathers and long decurved bill, originally found on all of the main Hawaiian Islands. Both of these birds are important pollinators within the forest, transferring pollen from flower to flower as they feed on nectar. Additionally, the Honeycreepers include several different insect feeders including the ʻAmakihi (*Chlorodrepanis virens*), the second most common of the native forest birds, as well as the very unique ʻAkiapōlaʻau (*Hemignathus wilsoni*), which has a long and decurved upper bill and a stout lower bill that is used to hammer the bark like a Woodpecker as it searches for insects. Other native birds shown in Photo 1.11, but not in the Honeycreeper family, are the ʻElepaio (*Chasiempis sandwichensis*), which catches insects in mid-air, and the ʻŌmaʻo (*Myadestes obscurus*), a fruit eating species that serves an important role in dispersing seeds that it ingests. Unfortunately, many of the Hawaiian birds are now extinct or very rare, falling victim to habitat loss, introduced avian disease, predation from introduced rats and cats, as well as competition from other alien bird species. The Palila (*Loxioides bailleui*), which is highly dependent on the subalpine dry forest, and the ʻAkiapolaʻau, depicted in Photo 1.11 are very rare and listed as Endangered Species by the US Fish and Wildlife Service. Two other birds also shown in that figure, the Poʻo-uli (*Melamprosops phaeosoma*) and the ʻOʻoʻaʻa (*Moho braccatus*), have gone extinct over the past 20 years.

Equally spectacular is Hawaiʻi's unique invertebrate diversity. Terrestrial invertebrate species include groups such as insects, spiders, and mites, as well as snails and slugs. It is estimated that the Hawaiian Islands have over 10,000 native invertebrate species and over 95% of them are endemic. Hawaiian rainforests are particularly rich habitats for invertebrates and ʻŌhiʻa itself has many species that are closely associated with it. These include the two-lined ʻŌhiʻa borer *Plagithmysus bilineatus*

Photo 1.12. Two-lined 'Ōhi'a borer (*Plagithmysus bilineatus*). This endemic beetle lays its eggs exclusively under the bark of dying 'Ōhi'a trees and the larvae feed on the decaying wood until they emerge as adults. Photograph by Karl Magnacca.

Photo 1.13. The Hawaiian predatory caterpilars in the genus *Eupithecia* are truly unique in the world. As adults they are small and relatively inconspicuous moths. However, the larvae are active predators that cling to the edge of a leaf and wait for an insect, such as a fly, to come close, whereupon they quickly grab it in their massive "forearms" and eat their fill. Photo by William P. Mull. © Bishop Museum, used with permission.

Photo 1.14. Hawai'i only has two native butterfly species and one of them is the spectacular Kamehameha butterfly (*Vanessa tameamea*), also known as Pulelehua. The adults are frequently seen in the understory of the rainforest feeding on the nectar of flowers while their caterpillars feed on the leaves of the native Mamake (*Pipturus albidus*), a shrub in the nettle family. Photo by William P. Mull. © Bishop Museum, used with permission.

(Photo 1.12), the predatory caterpillar *Eupithecia spp.* (Photo 1.13), and the beautiful Kamehameha butterfly (*Vanessa tameamea*) (Photo 1.14). Invertebrates have many significant roles within forest ecosystems, including pollination, decomposition of dead and dying plant material, as well as an important food resource for the native birds.

Roots of ʻŌhiʻa provide the resource for the unique Hawaiian lava tube ecosystem (see Frank Howarth 1981). Although we generally visualize ʻŌhiʻa as a sapling or tree, a large amount of its total biomass is found below ground as roots. It is extremely difficult to see this part of the tree but we get occasional glimpses into this below-ground component in lava tube

Photo 1.15. ʻŌhiʻa roots penetrating into a lava tube from a flow in the Puna District on the island of Hawaiʻi. Photo by J. Jacobi.

skylights (Photo 1.15). Lava tubes are formed when lava flows down the slope of a volcano like a river. While the surface cools, the flow may continue moving as a thick liquid below ground. When the eruption feeding this flow stops, the lava drains out of this underground channel leaving a hollow tube (Photo 1.16). The ʻŌhiʻa tree roots that penetrate this underground area are the primary source of organic material for an extremely unique ecosystem that was discovered by Frank Howarth and colleagues at the B. P. Bishop Museum in the early 1970s and includes a suite of cave adapted insects and spiders that are only found in Hawaiʻi.

Photo 1.16. View into a very hot and flowing lava tube coming from the Mauna Ulu vent on the flank of Kīlauea volcano in 1972. Photo by J. Jacobi.

Is there a climax forest in Hawai'i ?

A 200 year old forest is definitely mature, but it is not a climax forest. 'Ōhi'a lehua is a pioneer species, colonizing after canopy disturbances, such as the 1790 Kīlauea explosion that affected the crater rim area here. 'Ōhi'a lehua can be characterized as a "come-back species" but not as a typical climax species. Climax tree species of the sort found elsewhere in the tropics are, as yet, missing in all of Hawai'i's native rainforests. They never arrived here naturally. Climax species typically are shade-tolerant, slow growing trees that eventually overtop the pioneer species. When the first mature 'Ōhi'a canopy trees age and eventually senesce (get old) and die, they become replaced, and the next generation is again dominated by 'Ōhi'a lehua. This mature rainforest along thc Crater Rim Road is only a first generation forest following a catastrophic volcanic explosion of Kīlauea volcano that happened in 1790.

After the first generation 'Ōhi'a canopy trees die and are replaced, a second and then third generation forest develops. This turnover process coincides with aging and weathering of the volcanic substrate. The first lava rock or cinder-ash sub-

A CLIMAX FOREST, by definition, is dominated by late-successional tree species that become established as shade-tolerant trees in the undergrowth of a pioneer or early-successional forest. These late-successional (or climax) species eventually overtop the early-successional (or pioneer) species. This process of succession then results in a climax forest, whereby the pioneer species may be eliminated or reduced to a minor position, such as only surviving in canopy gaps of the dominating climax species or as epiphytes, plants perched high up in the branches of canopy trees. Climax forests in this sense occur only in regions with diverse tree floras, such as found in the continental tropics. This is the important difference in Hawai'i, which has a rather restricted indigenous tree flora due to its isolation in the Pacific.

strate represents a storehouse of mostly unavailable soil nutrients. These become available as time goes by with weathering of the stony substrate which results in the soil formation process. Cinder-ash overlying the clinker-like a'ā lava breaks down much faster than the pavement-like pāhoehoe lava. Nitrogen, which is essential for plants to grow, becomes more and more available through breakdown of organic matter on the forest floor and through fixation by bacteria of the nitrogen gas N_2, a resource of which there is plenty (almost 80%) in the air that we breathe, together with the oxygen contributed by the plants.

There are old growth forests

When the 'Ōhi'a rainforest has renewed itself after the third generation, an "old-growth" forest can get established (Photo 1.17 and 1.18). By that time the soil substrate may be 1,000 years old and relatively rich in available soil nutrients. The many trees and tree ferns that have died over that time span have enriched the nutrient capacity and improved the structure of the soil. The forest floor is strewn with decaying logs on which many native ferns, mosses, forbs, shrubs and tree seedlings can find favorable microhabitats.

A second endemic pioneer species, Koa (*Acacia koa*), may also join 'Ōhi'a in the tree canopy. A small remnant of such old-

OLD-GROWTH FORESTS typically contain tree species in several age states and size groups. Their main characteristic is the presence of very tall, large-diameter trees that are obviously much older than normally mature trees of the same species. When such big trees become senescent and thereafter break down, usually from root rot, they create individual tree fall gaps. As a rule such gaps are not very large. In Hawaiian old-growth forests, tree gaps are usually filled with shade-adapted plants, including the Hāpu'u tree ferns, many other ferns and shrubs such as pictured in this chapter. Larger gaps may be filled by small cohorts of 'Ōhi'a lehua.

Photo 1.17. Huge ʻŌhiʻa lehua trees like this one are found in an old-growth forest segment just north of the upper part of the old Volcano Road behind the Niaulani Cultural Center.

Photo 1.18. Looking upward through the understory along the trunk of the large ʻŌhiʻa tree in the forest on Photo 1.17 near Volcano. Hāpuʻu tree fern fronds grow up to 5 m high in this forest.

growth forest can be seen behind the Niaulani Cultural Center at the old Volcano Highway. What is striking here, besides noting the large-diameter and tall 'Ōhi'a trees associated with a few similarly large Koa trees, is the fact that several of the big 'Ōhi'a trees have rather large stilt roots revealing their epiphytic beginning. This means they started as seedlings on fallen logs of a former generation of 'Ōhi'a trees or on tree fern trunks. Their former support logs have mostly disappeared from decay, and the trees that germinated on the logs sent their roots down as aerial roots to be planted firmly in the mineral soil (Photo 1.19). These old-growth forests, such as found here at Niaulani, and also in the Ōla'a Tract and the Kīlauea Forest (Photo 1.20, 1.21, and 1.22), can be considered representing the "Hawaiian climax forest" with its tall growing (over 25 m) 'Ōhi'a and Koa trees. However, both species are shade-intolerant pioneer trees and thus are not true climax species as found elsewhere. Canopy gaps are typical for old-growth forests. Here are two profile

KOA (*Acacia koa*) is the second most widely occurring endemic forest tree in Hawai'i. It is, however, absent in most of the wetter (>3500 mm [>140 in] mean annual rainfall) rainforests. As a rainforest canopy tree mixed with 'Ōhi'a lehua, it is locally restricted to a lower belt on windward Mauna Kea and an upper belt on Hualālai, Mauna Kea and Mauna Loa volcanoes, here forming "old-growth forests." These Koa/'Ōhi'a mixed forests can be considered "Hawaiian climax forests" not to be confused with conventional "climax forests," (see box on p. 13). Koa is also a pioneer species, but usually does not colonize an area until after an 'Ōhi'a forest has become established. Very little is known about the distribution of Koa seeds. However, lots of Koa seed can be found distributed around the crown perimeter of mature trees. Scarifying (digging up) of the mineral soil around such trees usually results in the emergence of numerous seedlings. But in the broader rainforest territory dominated by 'Ōhi'a lehua, no Koa seedlings emerge on soil being scarified. It is possible that a former seed distribution agent has become extinct or that Koa seed is more widely dispersed only during unique storm events.

diagrams showing canopy gaps in the Kīlauea old-growth forest (Fig. 1.1, 1.2 and 1.3). The stand profiles are based on 6 x 20 m (19.7 x 65.6 ft) strip plots which were used for collecting vegetation data during the International Biological Program (IBP) research project (Photo 1.23, 1.24, and 1.25).

Photo 1.19. Lower part of same tree with moss covered stilt roots. The stilt roots indicate that the tree had an epiphytic beginning, meaning that it began as seedling on a tree fern trunk or decaying tree trunk and not on the mineral soil. As the tree grew up it became reestablished in the mineral soil via its stilt roots.

IBP is an acronym standing for **I**nternational **B**iological **P**rogram. The IBP was an international initiative to bring ecosystem researchers together working in teams. The general objective was to study "the biological basis of productivity and human welfare." Fifty-eight nations took part in the IBP. The Hawai'i-IBP was the island component among eight mainland US/IBP biomes funded by the National Science Foundation from 1967 through 1974. The Hawai'i-IBP was active from 1971 until the final Synthesis Volume was published in 1981: *Island Ecosystems: Biological Organization in Selected Hawaiian Communities*. See Suggested Readings list at the end of this chapter.

Photo 1.20. Aerial view of a large old Koa (*Acacia koa*) tree growing above the Hāpu'u tree fern canopy in the lower section of the Kīlauea Forest on the island of Hawai'i. Photo by J. Jacobi.

Photo 1.21. View within the Kīlauea forest, a mixed *Koa-'Ōhi'a* rainforest with an understory of native shrubs, small trees (Naio [*Myoporum sandwicense*] in center of photo), and tree ferns. This is another well-preserved old-growth forest on the east slope of Mauna Loa near the National Park. It is an upper montane rainforest in elevation between 5000–6000 feet (1525–1830 m). An 80 hectare (200 acre) plot in this forest served as a major study site during the Hawai'i IBP in 1971–1981. For location, see map, Fig. 1.3.

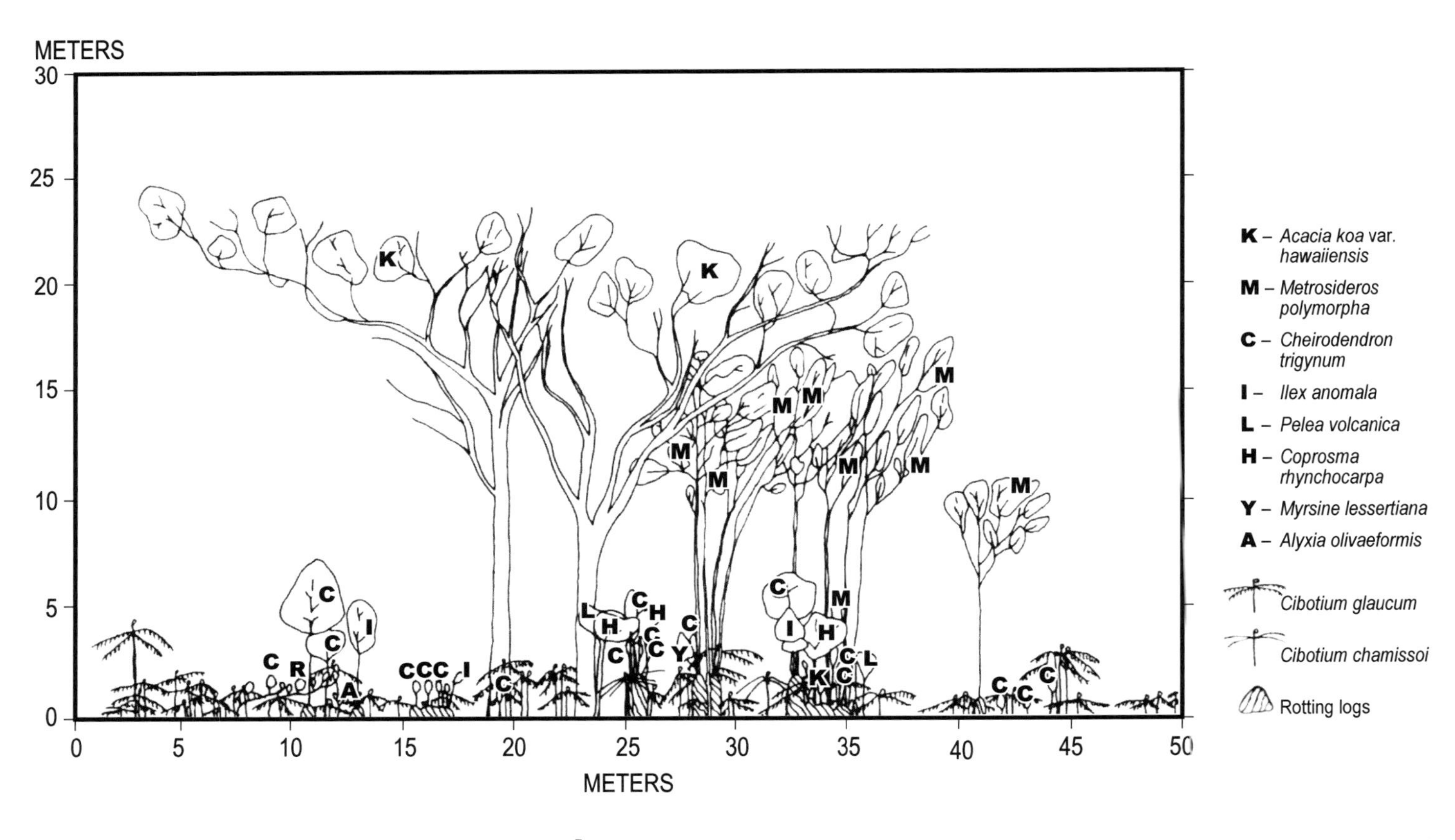

Fig. 1.1. On this stand profile 25 m tall Koa and 20 m tall ʻŌhiʻa old-growth survivors show canopy gaps on both sides. The gaps are filled with six sub-canopy tree species and two species of Hāpuʻu tree ferns. Drawing true to measurements by Ranjit Cooray.

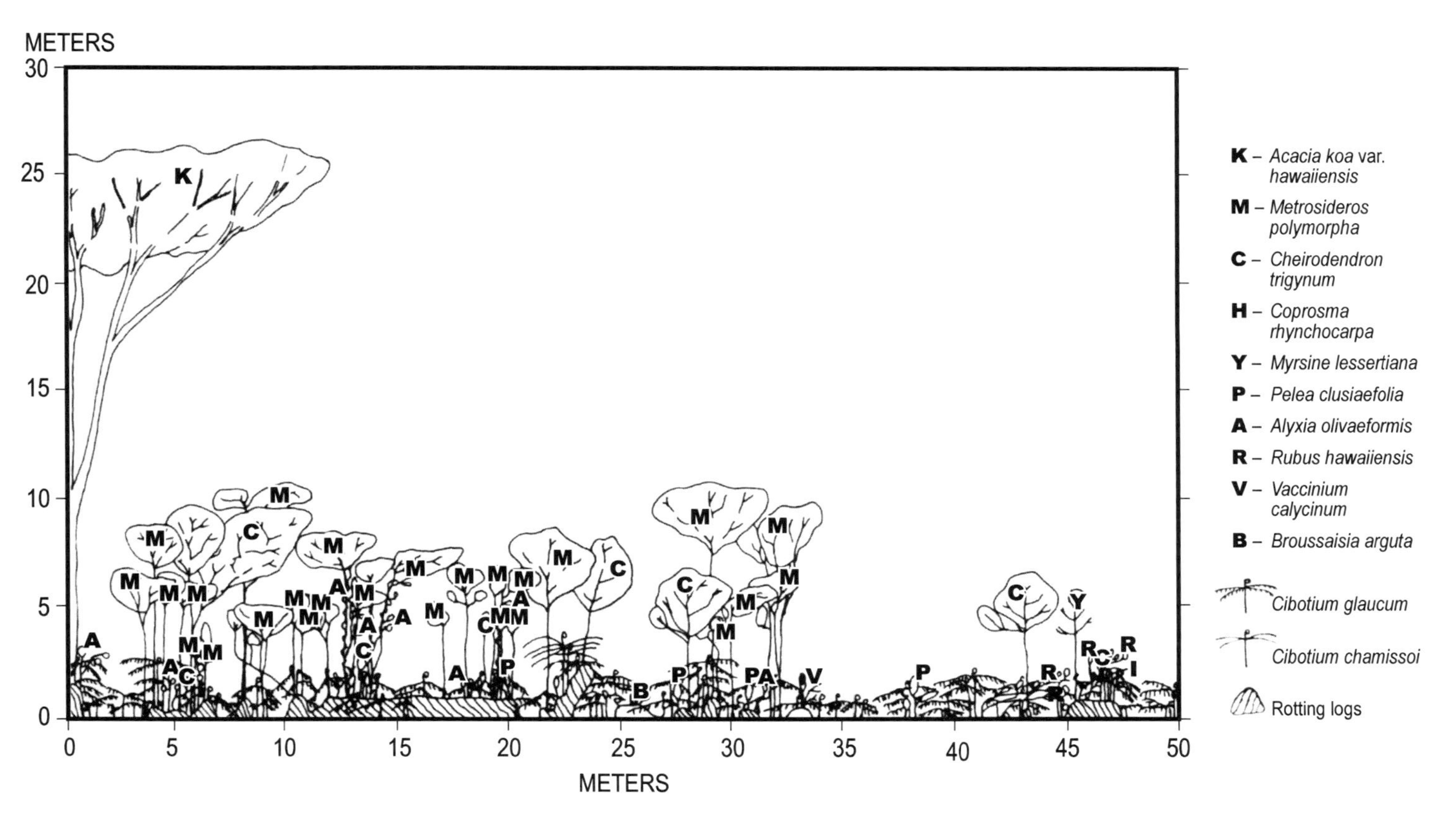

Fig. 1.2. Here a larger gap in the same old-growth forest supports a young cohort of ‘Ōhi‘a trees together with an assortment of ferns, shrubs, and potential sub-canopy trees. Drawing true to measurements by Ranjit Cooray.

A structurally simple forest system

In spite of additional species of plants, the general impression is a two- or three-species forest ecosystem, on account of the prevalence of the ʻŌhiʻa lehua trees leading the canopy layer generally at 12–15 m height and the tree ferns dominating in the undergrowth layer reaching up to 5 m height. A third prevalent species, the mat-forming Uluhe fern, occurs wherever the ʻŌhiʻa trees grow in more open stands or grow to smaller stature thereby allowing more light to reach the forest floor, which favors the Uluhe fern. The three kinds of plants, the ʻŌhiʻa lehua tree, the Hāpuʻu tree fern and the mat-forming Uluhe fern, can

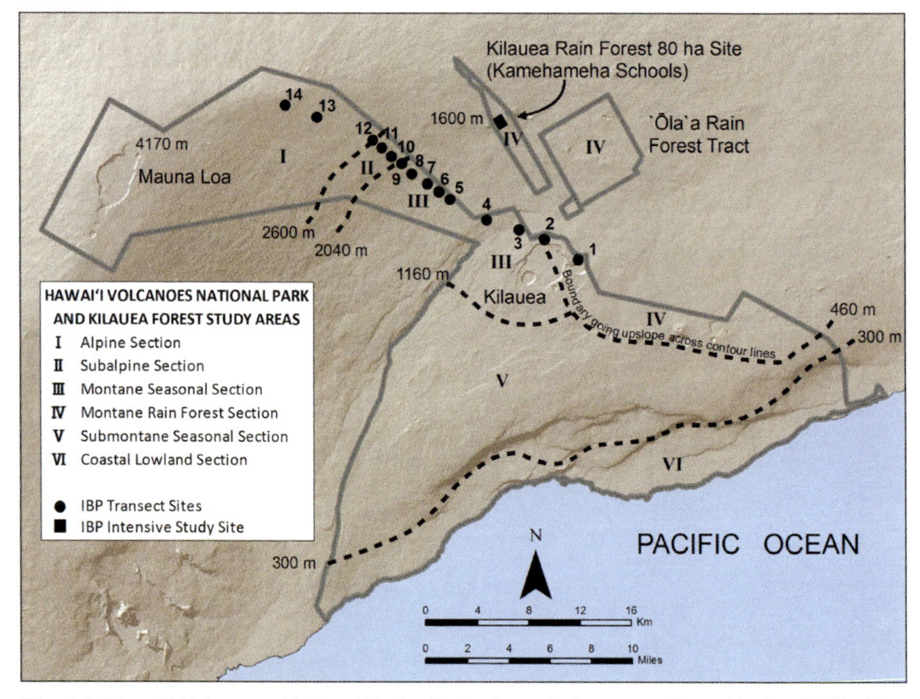

Fig. 1.3. Hawaiʻi Volcanoes National Park with the boundaries as configured in the 1970s. The map shows the 14 IBP transect sites along the Mauna Loa Transect. The IBP sites range from the rainforest section near the summit of Kīlauea volcano, mesic (moist) sites up the eastern slope of Mauna Loa, and into the dry alpine habitat above 2600 m (8530 ft) elevation. Also shown is the 80 ha (200 acre) Kīlauea Rainforest Site, north of the main transect, which is where Photos 1.20 through 1.23 were taken, as well as the areas depicted in Figures 1.1 and 1.2. Photos 1.24 and 1.25 were taken in ʻŌlaʻa Tract.

Photo 1.22 (above). Jim Jacobi counting tree seedlings on decaying tree trunk in the old-growth Koa-'Ōhi'a forest, the Kīlauea forest 80 ha (200 acre) IBP plot, 1974.

Photo 1.23. Ranjit Cooray standing next to a young Koa sapling growing on an overturned root collar in the Kīlauea forest IBP plot. The young Koa tree is growing in a canopy gap and off the ground, away from feral pig disturbance.

Photo 1.24. Close-up of endemic Loulu palm (*Pritchardia* sp.) with Nadarajah Balakrishnan climbing to collect fruit samples in 1978.

Photo 1.25. A senescing (declining) ʻŌhiʻa lehua old-growth forest in the Ōlaʻa Tract of Hawaiʻi Volcanoes National Park with a group of vigorous Loulu palms (the endemic *Pritchardia beccariana*) and Hāpuʻu tree ferns growing in the understory. The introduced Broomsedge is seen in the foreground along the side of Wright Road north of Volcano Village.

be called *"Keystone Species"* in the Hawaiian rainforest on account of their structural prevalence and their functional importance. Rainforests with Koa in the canopy are confined usually to less wet portions of the rainforest, which are not as prevalent.

A forest formed in isolation

What is so remarkable about Hawai'i's 'Ōhi'a rainforest is the fact that it is composed of only a few prevalent tree species that recur in similar combinations as dominants across the islands. Tropical rainforests elsewhere in similar environments contain far more tree species in their canopy (Mueller-Dombois & Fosberg 1998). The reason for this scarcity of tree species is the isolation of the Hawaiian Islands (Fig. 1.4). They arose as volcanic islands in the north-central Pacific far away (more than 3,200 km or 2,000 miles) from any other landmass and they never were connected to any other landmass in their historic past. This isolation factor is a major constraint in the Hawaiian rainforest. It is actually a wonder that native forests did arise here at all, because the ancestral species had to come from somewhere far away. They did not arise in the middle of the ocean as did the volcanic substrate on which they became established. Yet the Hawaiian rainforest evolved clearly into a self-perpetuating and fully functional ecosystem. But as a tropical forest it is definitely unique, because here, isolation has brought about a species assembly in the Hawaiian rainforest that can be called ***"indiginally impoverished but endemically enriched"*** as compared to other tropical rainforests in the World. While this is structurally obvious as described above, it is also functionally significant as we will see when we consider the topic of rainforest dynamics in Hawai'i (later in this book).

The Hawaiian rainforest, a mosaic of cohort stands

The 'Ōhi'a/tree fern forest at the entrance to Hawai'i Volcanoes National Park and along the Crater Rim Road was most-

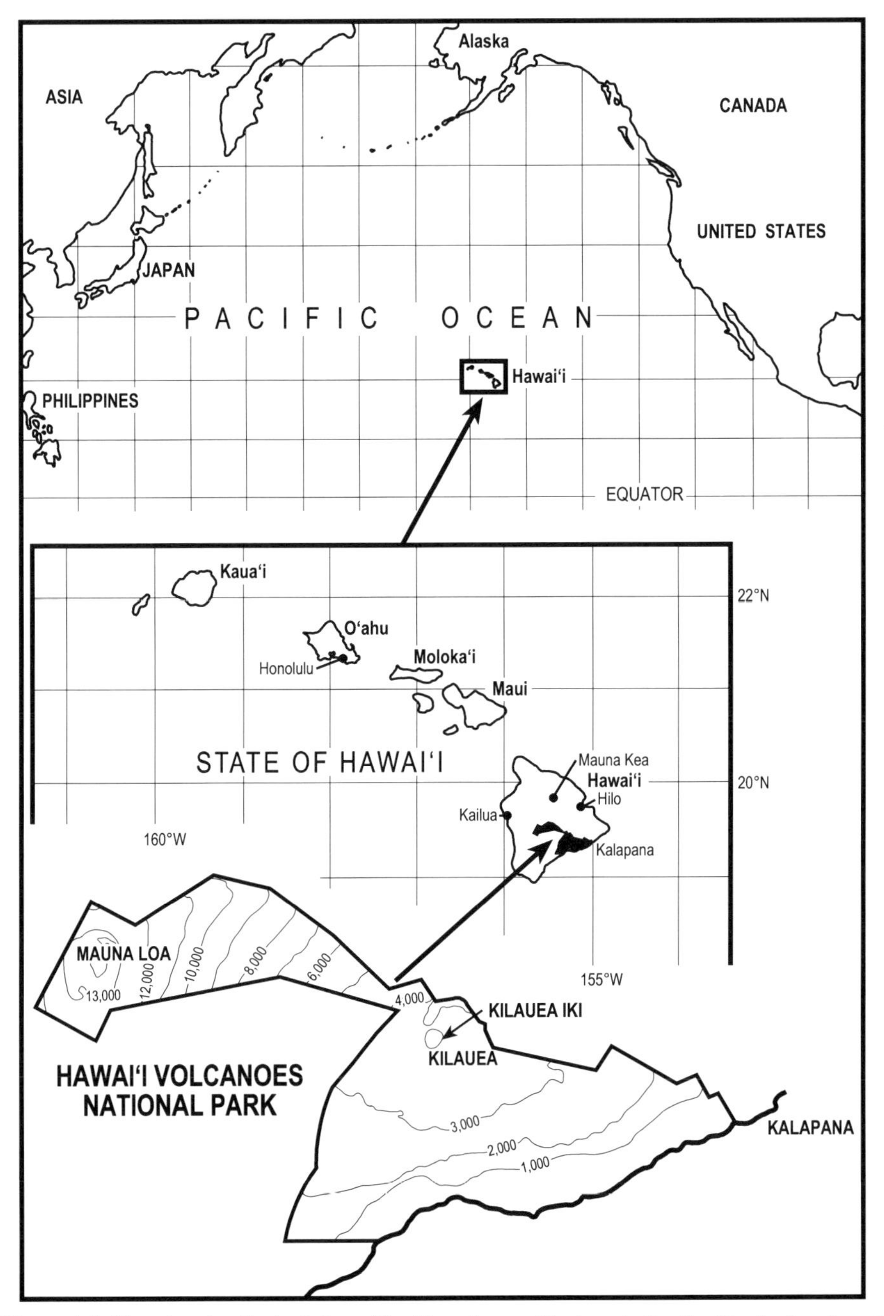

Figure 1.4. Map showing the isolation of the Hawaiian archipelago in the Pacific, the location of Hawai'i Volcanoes National Park on the Big Island and the two active volcanoes, Kīlauea and Mauna Loa. The native rainforest extends from the 4,000 ft (1,200 m) elevation mark all the way east to Kalapana, and northward across the east slope of Mauna Kea midway between the summit and down to Hilo.

ly destroyed after the huge explosive Kīlauea eruption which occurred in 1790. The ʻŌhiʻa lehua forest grew back in less than 200 years and recovered to what is now a mature, first generation rainforest. Many of the forest stand segments here now look almost as if the trees were planted, perhaps about 200 years ago. The uniform tree structure indicates that most of the trees here grew up together in form of generation stands or cohorts, after that destructive explosion. Thus in terms of stand demography, we can characterize this rainforest as composed of ***"cohort forest stands."***

COHORT FOREST STANDS are formed by individuals of a population that are born over a certain limited time span (in ʻŌhiʻa stands within a span of about 50 years), and that grow up together through their life cycle, experiencing the same favorable and stressful periods during the history of their demographic phases of life.

An aerial view of the vegetation on the northeast flank of Mauna Loa (Photo 1.26), shows a mosaic of ʻŌhiʻa cohort stands associated with lava flows. In the foreground at lower right is a young forest on the 1855 flow. Next to it is an older cohort forest in dieback condition. The darker segments are rainforests with foliated ʻŌhiʻa canopies belonging to different mature cohorts.

The aerial view on Mauna Kea (Photo 1.27) also shows segments of rainforests representing ʻŌhiʻa cohorts. But in contrast to those on Mauna Loa, they seem to be more associated with topography. These patterns may have started with lava flows or volcanic ash deposits but have subsequently undergone geomorphological development. In both cases there is an underlying history of substrate disturbances that resulted in vegetation recovery with the ʻŌhiʻa tree forming separate canopy cohorts.

Photo 1.26. Aerial view of various ʻŌhiʻa cohorts on different aged lava flows on the northeast flank of Mauna Loa. The young forest at the lower right is growing on the 1855 lava flow. Next to it is an older ʻŌhiʻa forest which was in dieback condition when the photo was taken in the mid-1970s. The darker forest segments in the background are healthy ʻŌhiʻa forest stands belonging to different mature cohorts. Photo by J. Jacobi.

Photo 1.27. Aerial view of a cohort mosaic on the east slope of Mauna Kea showing different cohorts of ʻŌhiʻa forest. In the foreground the open ʻŌhiʻa forest is growing in a matrix of Uluhe ferns; in the background is a tall, closed canopy ʻŌhiʻa forest with a mix of tree ferns and native shrubs in the understory. Photo by Rick Warshauer, April 2005.

These three structural characteristics: few keystone species, floristic impoverishment, and cohort stand structure are of great relevance to the functioning of the Hawaiian rainforest as will be explained in the following chapters.

Suggested Readings

Atkinson, I. A. E. (1970). Successional trends in the coastal and lowland forest of Mauna Loa and Kīlauea volcanoes, Hawaii. *Pacific Science* 24: 387–400.

Carlquist, S. (1980). *Hawaii, A Natural History. Geology, Climate, Native Flora and Fauna Above the Shoreline*. 2nd Edition. Honolulu: S. B. Printers for Pacific Tropical Botanical Gardens. 468 p.

Cuddihy, L. W. & Stone, C. P. (1990). *Alteration of Native Hawaiian Vegetation: Effects of Humans, Their Activities and Introductions*. Honolulu: University of Hawaiʻi, Cooperative National Park Resources Studies Unit. 138 p.

Culliney, J. L. (2006). *Islands in a Far Sea: The Fate of Nature in Hawaiʻi*. Revised edition. Honolulu: University of Hawaiʻi Press. 420 p.

Gerrish, G. & Mueller-Dombois, D. (1999). Measuring stem growth rates for determining age and cohort analysis of a tropical evergreen tree. *Pacific Science* 53: 418–429.

Hart, P. J. (2010). Tree growth and age in an ancient Hawaiian wet forest: vegetation dynamics at two spatial scales. *Journal of Tropical Ecology* 25:1–11.

Hatfield, J. S., Link, W. A., Dawson, D. K. & Lindquist, E. L. (1996). Coexistence and community structure of tropical trees in a Hawaiian montane rain forest. *Biotropica* 28: 746–758.

Howarth, F. G. 1981. Community structure and niche differentiation in Hawaiian lava tubes. In *Island Ecosystems: Biological Organization in Selected Hawaiian Communities*, ed. by D. Mueller-Dombois, K. W. Bridges, & H. L. Carson. Stroudsburg, PA: Hutchinson Ross Publishing Company. pp. 318–336.

Juvik, S. P. & Juvik, J. O. (1998). *Atlas of Hawaiʻi*. 3rd Edition. Honolulu: University of Hawaiʻi Press. 333 p.

Medeiros, A. C., Loope, L. L. & Anderson, S. J. (1993). Differential colonization by epiphytes on native (*Cibotium* spp.) and alien (*Cyathea cooperi*) tree ferns in a Hawaiian rain forest. *Selbyana* 14: 71–74.

Mueller-Dombois, D. & Fosberg, F. R. (1998). *Vegetation of the Tropical Pacific Islands*. New York: Springer-Verlag New York, Inc. 733p.

Mueller-Dombois, D., Bridges, K. W. & Carson, H. L. (eds.) (1981). *Island Ecosystems: Biological Organization in Selected Hawaiian Communities*. Stroudsburg, PA & Woods Hole, MA: Hutchinson Ross Publishing Company, US/IBP Synthesis Series 15. 583 p.

Pratt, T. K., Atkinson, C. T., Banko, P. C., Jacobi, J. D. & Woodworth, B. L. (eds.) (2009). *Conservation Biology of Hawaiian Forest Birds*. New Haven & London: Yale University Press. 707 p.

Rock, J. F. (1974). *Indigenous Trees of the Hawaiian Islands*. Reprint from 1913 edition. Lāwa'i, Kaua'i: Pacific Tropical Botanical Garden. 548 p.

Stone, C. P. & Scott, J. M. (1984). *Hawai'i's Terrestrial Ecosystems: Preservation and Management*. Honolulu: University of Hawai'i, Cooperative National Park Resources Studies Unit. 584 p.

Stone, C. P. & Pratt, L. W. (1994). *Hawai'i's Plants and Animals: Biological Sketches of Hawaii Volcanoes National Park*. Honolulu: Hawai'i Natural History Association, National Park Service and University of Hawai'i Cooperative National Park Resouces Studies Unit. Distributed by University of Hawai'i Press. 399 p.

Wagner, W. L., Herbst, D. R. & Sohmer, S. H. (1999). *Manual of the Flowering Plants of Hawai'i*. 2nd edition. Vol. 1 and Vol. 2. Honolulu: University of Hawai'i and Bishop Museum Press, Bishop Museum Special Publication 83. 1854 p.

Ziegler, A. C. (2002). *Hawaiian Natural History, Ecology, and Evolution*. University of Hawai'i Press, Honolulu. 478 p.

Chapter 2

Origin of ʻŌhiʻa Forest Among Volcanoes

ʻŌhiʻa lehua, the first tree on new volcanic surfaces

Volcanic **explosions** are a common phenomenon in Hawaiʻi Volcanoes National Park. They cause destruction of existing vegetation but also produce new volcanic surfaces. The more common surfaces are lava flows, less common are cinder surfaces and cinder cones. Both types are

Photo 2.1. Red hot lava erupting from Kīlauea Iki in 1959. Photo courtesy of National Park Service.

part of the "Devastation Area" in Hawai'i Volcanoes National Park that resulted from the 1959 explosion of Kīlauea Iki (Photo 2.1, 2.2, and 2.3).

Two kinds of lava, known as pāhoehoe and 'a'ā are the most common surfaces on the Kīlauea and Mauna Loa volcanoes. Pāhoehoe results from relatively hot and fast moving flows. When cooled, pāhoehoe looks like fragments of pavement (Photo 2.4 and 2.5). 'A'ā lava results usually from impediments or obstacles in a pāhoehoe flow, such as a suddenly encountered depression. The smooth surfaces cool somewhat and then change abruptly into clinker-like rocks, often of football size with sharp points all around. These clinker-like rocks move on top of a red-hot glowing core of lava. The solid lava core is usually a meter (3 feet) below its burden of loose clinker-like rocks (Photo 2.6). The 'a'ā flow then moves slowly forward like

Photo 2.2. The Pu'u Pua'i cinder cone that buried a rainforest during the Kīlauea Iki eruption in 1959. Photo taken in 1984.

Photo 2.3. The cinder strewn area south of Kīlauea Iki and the Puʻu Puaʻi cinder cone showing the dead trunks of trees killed during this eruption and live trees that survived in the background. A portion of the Devastation Trail is seen to the left of the picture. Photo taken in 1981.

Photo 2.4. The pāhoehoe flow of 1974 with small kīpuka and Garrett Smathers standing in the back. This flow originated from a fissure that cut through the forest that survived the 1959 Kīlauea Iki eruption and was extruded as "shelly pāhoehoe lava" over the 1959 cinder in the lower part of the Devastation Area. Photo taken in 1998.

Photo 2.5. A pāhoehoe lava flow inside Makaopuhi Crater which is in Hawai'i Volcanoes National Park. Former botany colleagues Ray Fosberg, Max Doty, and Charles Lamoureux are discussing where plant invasion would start. This is a massive pāhoehoe flow contrasting with the shelly flow on the previous photo. Photo taken one year after the Makaopuhi eruption in 1965.

Photo 2.6. The 1977 a'ā flow above Kalapana 10 years after it erupted. Ferns and 'Ōhi'a lehua seedlings are growing out of the fissures and the native lichen *Stereocaulon vulcani* covers most of the surface of the lava. Photo taken in 1987.

the chains on the wheels of a military tank or bulldozer, often burying the surface load of clinker-like rocks again in its forward movement (Photo 2.7). When cooled, ʻaʻā flows are a forbidding surface to walk on.

The first colonizers. When lava flows cover rainforest habitats, which we define as environments receiving at least 100 mm (4 in) of rainfall/month or greater than 1200–1500 mm (> 47–59 in) of rainfall/year, they initially look like a stone desert. But new vegetation develops within a few years. First one may notice the white-gray colored lichen *Stereocaulon vulcani* which covers the rock surfaces. When looking more closely, one may see plant life emerging among the rocks and fissures. The early arrivers are plants that propagate by spores, such as algae, mosses and ferns. The most conspicuous fern is the false ʻŌkupukupu (*Nephrolepis multiflora*), a small alien fern (Photo 2.8). Native

Photo 2.7. Massive ʻaʻā lava flow moving through a mature ʻŌhiʻa forest on the east flank of Mauna Loa volcano in 1984. The ʻŌhiʻa trees in this picture are over 10 m tall and the top of the flow is nearly twice that height. Photo by J. Jacobi.

seed plants arrive soon thereafter. Most regular among them are the native low-growing pioneer shrubs Kūpaoa (*Dubautia scabra*) (Photo 2.9), and 'Ōhelo 'ai (*Vaccinium reticulatum*) (Photo 2.10). At the same time 'Ōhi'a lehua seedlings can be seen. 'Ōhi'a is the first native tree to arrive on new volcanic surfaces, since seed source trees are typically nearby in untouched forest stands (Photo 2.11, 2.12, and 2.13).

Photo 2.8. False 'Ōkupukupu ferns (*Nephrolepis multiflora*) in a crevice of the 1955 Puna pāhoehoe flow. Again, most of the rock surfaces are covered by the native lichen *Stereocaulon vulcani*. Photo taken in 1965.

Photo 2.9. Kūpaoa (*Dubautia scabra*) an endemic low shrub that is colonizing the 1959 cinder habitat in the Devastation Area. In contrast to 'Ōhelo ai, this is a short-lived perennial shrub, which tends to senesce after 15—25 years. It then provides favorable microhabitats for other colonizers, including 'Ōhi'a lehua. Here, Bryce Decker points at two life phases, a flowering bush at the right and a bush beginning to senesce on the left. Photo taken in 1998.

Photo 2.10. 'Ōhelo 'ai (*Vaccinium reticulatum*), an endemic low-growing early colonizer is a pioneer shrub that can grow directly out of the cinder surface. Seeds of its red berries may be distributed by the endemic Nēnē goose (*Branta sandvicensis*) (Photo 2.15) or by wind when the fruits are dry. Photo taken in 1999.

Photo 2.11. A small 'Ōhi'a seedling, with its roots exposed, emerging from a crack on the "bathtub ring" of the 1959 Kīlauea Iki lava lake. Here, as in the following photo, 'Ōhi'a can grow independently, meaning it does not require supporting plants or organic matter as do many other colonizers. Photo taken in 1966.

Photo 2.12. A lone ʻŌhiʻa lehua seedling growing in a crevice on the south slope of the Pu'u Puaʻi cinder cone. Tree seedlings are rarely the first woody species colonizers on cinder surfaces except in cracks or crevices such as these. Photo taken in 1976.

Photo 2.13. Small ʻŌhiʻa sapling growing close to the same place as Photo 2.11. Although it is a young plant it is already able to flower. Photo taken in 1981.

Photo 2.14. ʻŌhiʻa flower with emerging buds hanging below. This is the pubescent variety (*Metrosideros polymorpha var. polymorpha*), which is common on new lava flows in drier areas and higher elevations than the glabrous or smooth variety (*M. polymorpha var. glaberrima*). Both varieties can often be found growing in the same forest.

Photo 2.15. A flock of endemic Nēnē geese (*Branta sandvicensis*) standing in an early successional habitat. Nēnē are fond of grass seeds and berries, such as the fruit of the endemic ‘Ōhelo ‘ai shrub shown in Photo 2.10. Nēnē are suspected to help restock their own fruit orchard via their droppings on new volcanic surfaces.

Microhabitats are of critical importance in early vegetation development. Rock surfaces are often soon occupied by the white-gray crustose lichen (*Stereocaulon vulcani*), that may grow slowly into branched forms, 1–3 cm (0.5–1.5 in) high. Most other pioneer plants colonize first in cracks or fissures among the rock surfaces. Fissures, where some small rock chips, dust, and bits of organic matter have accumulated from movement by wind or water, are favorable microhabitats.

When rainfall strikes a pāhoehoe surface the water is immediately channeled into an unequal distribution system, making some fissures wetter and others drier. Pāhoehoe plates pushed together (as shown on Photo 2.5) can provide an “inverted funnel effect.” The wetter microhabitats are favored by the pioneering plants. A uniform cinder surface is not easily invaded by ‘Ōhi‘a seedlings. Here low-growing pioneering shrubs, such as the endemic ‘Ōhelo ‘ai and Kūpaoa, are the best colonizers. They immediately develop a lateral rooting system beneath the cinder surface that catches rainwater over at least the area of their above-ground shoot system.

New volcanic surfaces dry up immediately after the rain stops. The ability to survive in such harsh environments depends on the plant's capacity to stretch its new roots quickly to zones away from the evaporating surface. Rain striking a new volcanic surface penetrates quickly, but it leaves a layer of film water on the rock particles that thins quickly and then becomes unavailable for plants. The depth of the evaporating surface zone depends in part on the intensity and duration of the evaporative power of the air following a rainy period. The texture of the new volcanic surface is even more critical. A coarse rock textured surface has practically no water holding capacity, except that a film of available water may adhere to fissures a meter (3 feet) or so down in the substrate. Fast root extension to that adhering film, away from the evaporation zone, is essential for early survival. Cinder and ash substrates, which are of finer rock texture also allow water to penetrate rapidly. But more water adheres to the much greater surface area of that finer textured material. For plants to survive, initial lateral extension of their root system is more critical than initial vertical extension. That difference in early rooting capacity may promote the low shrubs to be the first successful pioneers on cinder, while 'Ōhi'a is among the first woody plants on lava substrates because of their superior capacity for vertical root extension.

Seed availability and distribution—A seed source is just as important as a new rain-fed substrate to grow on. A mature 'Ōhi'a forest is remarkable in that there are always some trees in flower. Usually, more trees flower in the cool season, but there are no distinct seasons without seeds being available. This is of critical importance for this tree species. It guarantees survival in the unpredictable volcanic environment, where a catastrophic disturbance can happen any time of the year. A tree species with distinct flowering seasons interrupted by non-flowering seasons would not survive in such unpredictable environment. Thus, ʻŌhiʻa lehua is well adapted to grow in a volcanic environment where catastrophic disturbances can happen unexpectedly at any time. The capacity of ʻŌhiʻa for continuous recolonization is an essential survival feature.

Seed production alone is not enough for the next generation to take hold. The seed distribution system must be wide-reaching. Such widespread distribution of ʻŌhiʻa lehua seeds has been verified by seed trap studies (Drake 1992). It can also be witnessed wherever a generation of ʻŌhiʻa seedlings invades and becomes established on a huge lava field. This feature is made possible by the small, light-weight seeds that are easily distributed by wind (Photo 2.16).

Photo 2.16. Open seed capsules showing seeds of ʻŌhiʻa lehua. The seeds are only 5 mm long by 1 mm wide and are very light and easily blown by the wind over long distances of 100–1000 meters and more (exceeding a mile). Photo by Alvin Yoshinaga.

Succession to mature rainforest

About 15 km east of Hawaiʻi Volcanoes National Park, in the humid lowland is the Kamāʻili lava flow complex. There, a 1955 ʻaʻā flow crosses the road from Pāhoa to Kalapana at 305 m (1,000 feet) elevation. An early picture (Photo 2.17) from 1971 shows a number of ʻŌhiʻa seedlings emerging among the rocks covered with the white-gray growth of the lichen *Stereocaulon vulcani*. Note the kīpuka in the background. The next picture (Photo 2.18) was taken from the same spot 14 years later in 1985. The seedlings have grown into ʻŌhiʻa saplings, and a scattering of ferns and pioneer shrubs becomes conspicuous. The two following pictures (Photo 2.19 and 2.20) provide for a wider view. One may get the impression that the ʻŌhiʻa tree saplings have been planted there.

Photo 2.17. 1971 photo of 1955 Kamāili ʻaʻā flow in the Puna district of the island of Hawaiʻi. The ʻŌhiʻa forest stand in the background is growing in a kīpuka, an older lava surface that was surrounded by the younger lava flow.

KĪPUKA is a Hawaiian word for an island of vegetation within a lava flow, also referred to as "an oasis within a lava bed" in the Hawaiian dictionary by Pukui and Elbert (1986). A kīpuka can also be an island on an ash dune surrounded by a flow or a depression, a hole or gap in a forest. Usually, kīpuka are best seen as remnants of an earlier vegetation isolated within a new volcanic substrate as the kīpuka on Photo 2.17 and 2.18. Such kīpuka can serve as a seed source for plant invasion. The word kīpuka has been adopted as a scientific term in vulcanology worldwide.

Photo 2.18. The same Kamā'ili 'a'ā flow area shown in Photo 2.17 fourteen years later. Photo taken in 1985. The vegetation on the 1955 lava flow has grown much taller but the forest in the background kīpuka has not changed very much.

Photo 2.19. Another view of the pioneer 'Ōhi'a colonizing the Kama'ili 'a'ā lava flow in 1985.

Photo 2.20. More young 'Ōhi'a trees growing on the Kama'ili 'a'ā lava flow in 1985.

Photo 2.21 shows a young sapling cohort of 'Ōhi'a in front of a cohort of an early juvenile 'Ōhi'a forest.

Early succession occurs in the undergrowth. Photo 2.22 shows the successional change in the undergrowth of a more open grown juvenile rainforest stand, and Photo 2.23 another view inside that juvenile cohort forest, whose age can be estimated as about 100 years. There are many such juvenile stands, for example in Fern Acres subdivision, along the South Kūlani Road going south from Mountain View on Highway 11, or on the 1881 lava flow at the lower part of the Saddle Road. The change in undergrowth from broad-area *Stereocaulon* lichen cover with low-growing pioneer shrubs to broad-area Uluhe fern cover is evident in many places where differently aged lava flows are found. It shows that succession is taking place primarily in

Photo 2.21. Another view of the Kamā'ili 'a'ā flow in 1985 showing the different aged cohorts of 'Ōhi'a on the young lava flow versus the older substrate in the background.

Photo 2.22. A juvenile Ōhi'a rainforest with Lani Stemrnermann standing next to the dense undergrowth of the mat-forming Uluhe fern. At this stage of succession of the forest the Uluhe effectively blocks the light reaching the surface of the ground which prevents new Ōhi'a seedlings from joining the already established tree cohort.

Photo 2.23. A closer view of a young 'Ōhi'a forest showing the dense Uluhe fern completely dominating the understory. Photo taken in 1985.

the undergrowth, while the first-generation ‘Ōhi‘a forest develops to maturity.

Photo 2.24 shows a 200 year old mature cohort forest that belongs to the Kamā‘ili lava complex. Unfortunately, this forest is unprotected and unmanaged. It harbors only a few scattered tree ferns under its canopy and is fringed at the road by weed shrubs and trees, such as Strawberry guava (*Psidium cattleianum*). Tree ferns are sought after as a resource for gardening purposes, such as growing orchids. While ‘Ōhi‘a tree generations grow up on lava flows (Photo 2.25), succession occurs in the undergrowth from lichen cover with low growing shrubs to the light-loving Uluhe fern to the shade-tolerant Hāpu‘u tree ferns, arborescent shrubs and sub-canopy trees under the ‘Ōhi‘a canopy cohort.

Photo 2.24. A 200 year-old ‘Ōhi‘a forest also in the Puna district at 305 m (1000 ft) elevation. In this older forest the trees are much taller and the tree canopy is starting to dominate the vegetation, eliminating most of the light gaps in the understory which has caused the Uluhe fern to likewise disappear. Photo taken in 1985.

Photo 2.25. A view of the mosaic of different aged 'Ōhi'a forests growing on lava flows ranging in age from less than 150 years to over 3,000 years. The forest in the foreground experienced canopy dieback in the late 1960s to the 1970s. Photo by J. Jacobi.

Suggested Readings

Drake, D. R. (1992). Seed dispersal of *Metrosideros polymorpha* (Myrtaceae): A pioneer tree of Hawaiian lava flows. *American Journal of Botany* 79(11): 1224–1228.

Drake D. R. & Mueller-Dombois, D. (1993). Population development of rain forest trees on a chronosequence of Hawaiian lava flows. *Ecology* 74: 1012–1019.

Hotchkiss, S., Vitousek, P. M., Chadwick, O. A. & Price, J. P. (2000). Climate cycles, geomorphological change, and the interpretation of soil and ecosystem development. *Ecosystems* 3: 522–533.

Kitayama, K., Mueller-Dombois, D. & Vitousek, P. M. (1995). Primary succession of Hawaiian montane rain forest on a chronosequence of eight lava flows. *Journal of Vegetation Science* 6: 211–222.

Mueller-Dombois, D. (1996). Geomorphological evolution of stream ecosystems and rain forest dynamics in Hawaiʻi. In *Will Stream Restoration Benefit Freshwater, Estuarine, and Marine Fisheries?* ed. by W. S. Devick. Honolulu: Hawaiʻi Department of Land and Natural Resources. pp. 7–29.

Percy, D. M., Garver, A. M., Wagner, W. L., James, H. F., Cunningham, C. W., Miller, S. E. & Fleischer, R. C. (2008). Progressive island colonization and ancient origin of Hawaiian *Metrosideros* (Myrtaceae). *Proc. Royal Soc.* B 275: 1479–1490.

Pukui, M. K. & Elbert S. H. (1986). *Hawaiian Dictionary*. Rev. and enlarged ed. Honolulu: University of Hawaiʻi Press.

Smathers, A. G. & Mueller-Dombois, D. (2007). *Hawaiʻi, The Fires of Life.* Honolulu: Mutual Publishing. 142 p.

Stemmermann, L. (1983). Ecological studies of Hawaiian *Metrosideros* in a successional context. *Pacific Science* 37: 361–373.

Chapter 3

The ʻŌhiʻa Rainforest in a Landscape Perspective

Climate diagram maps to identify rainforest territory

As the term rainforest indicates, the amount of rainfall and its month-to-month reliability is a major determinant for rainforest to develop and occupy a landscape. The other major determinant is the forest-plant assemblage as discussed in Chapter 1.

First, let's focus on rainfall: On the map showing the Hawaiian archipelago from Niʻihau in the upper left to the Big Island of Hawaiʻi at the lower right, there are four climate diagrams indicating rainforest (Fig. 3.1). They refer to locations on the windward side of Kauaʻi (Mount Waiʻaleʻale 1,569 m, or 5,273 ft, above sea level), Oʻahu (Kāneʻohe 60 m, 200 ft), Maui (Keʻanae 298 m, 983 ft) and Hilo on the Big Island (12 m, 40 ft). These climate diagrams stand out by showing solid black fields in the upper part of each diagram. The black fields begin at a horizontal line, which is drawn at 100 mm (4 in) with reference to the right-hand ordinate (see the Mt. Waiʻaleʻale diagram where the line is marked 100 mm on the right-hand ordinate). Above the 100 mm horizontal line, rainfall is scaled in 200 mm (8 inch) intervals, while below the 100 mm line, rainfall is scaled in 20 mm (0.8 inch) intervals. This change in scale allows high rainfall amounts to be accommodated in a compact diagram. There is an important biological reason for the 100 mm (4 inch) line. Any location where the average month-to-month rainfall exceeds 100 mm can be considered rainforest territory. This is an acceptable definition for a climate that supports rainforest.

The month-to-month rainfall curve is the zigzag line at the top of these diagrams. This line shows the monthly variation of rainfall. In Hilo it averages about 300 mm/month (about 12 in/month), at Waiʻaleʻale about 1,000 mm/month (40 in/month). The latter is arguably one of the wettest spots in the World with average rainfall exceeding 12.3 meters/year (41 ft/yr).

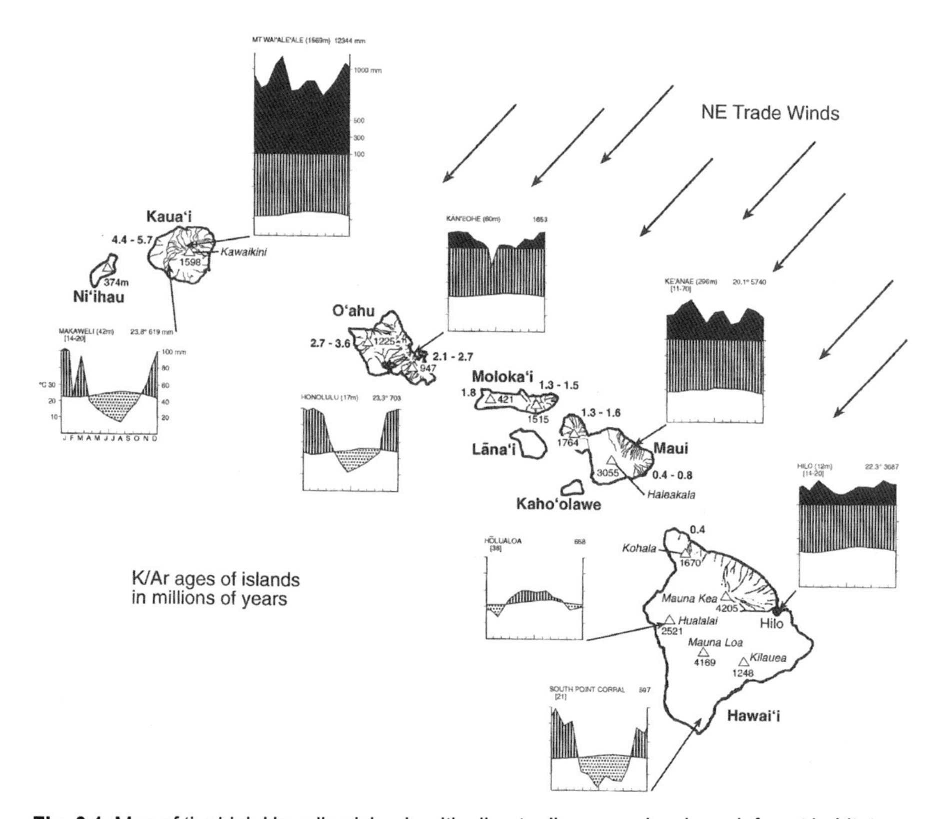

Fig. 3.1. Map of the high Hawaiian islands with climate diagrams showing rainforest habitats on the windward sides of the islands, indicated by the black sections at the top of the graphs. On the leeward sides of the islands the climate is much drier, showing moisture deficits for parts of the year, indicated by the hatched sections below the temperature line on each graph. The small wiggly black lines starting north of Hilo depict streams. Note the near complete absence of streams on the two youngest volcanic mountains, Mauna Loa and Kīlauea. Elevation of each mountain is given as meters above sea level and island ages in millions of years.

Here, it is so wet that ʻŌhiʻa only grows as dwarf trees in bogs. The Hawaiian flora lacks trees that can grow tall in such a wet place. The other three diagrams, including Kāneʻohe on Oʻahu show typical rainforest climate.

Next, let's focus on temperature: There is a smooth, almost normal curve in each climate diagram. This is the average month-to-month temperature. It is oriented to the left ordinate, where the air temperature scale is entered at 10 °C intervals. The mean air temperature in the Hawaiian lowlands is typically above 20 °C (68 °F) as shown on the Hilo and Honolulu diagrams. Upslope the temperature decreases according to the lapse rate. The average monthly temperature at Waiʻaleʻale on Kauaʻi at 1,569 m (5,273 ft) altitude hovers just above 10 °C (50 °F) through the year. This is certainly, a cold place in a tropical island. But it is still a tropical environment. The annual temperature curve is the evidence for that. It differs only little between summer and winter. It shows that September is the warmest month and February the coolest at any elevation. This difference of the warmest and coolest month is only about 5–6 °C (41–43 °F).

Wet versus dry seasons: Now looking at the Honolulu diagram, we can see that the rainfall curve undercuts the temperature curve. The undercutting begins in May and ends in October. This means there is usually a 5–6 month long dry season on the leeward side in Honolulu. In fact, at the scale of the two y-axes where 10 °C on the left y-axis is made parallel to 20 mm precipitation on the right ordinate, there is normally a rather severe dry season. This severe dryness can be considered a drought season. During such a season rain fed annual plants die and survive only in form of seeds.

The climate diagrams are useful predictors of the kinds of vegetation to be expected around the climatic stations from which the data are extracted. They do not reveal the vegetation itself.

Since we are concerned here primarily with the native 'Ōhi'a rainforest on Hawai'i Island, let us focus next on the climate diagram map of this Island (Fig. 3.2). This Big Island map shows 12 climate diagrams. We recognize the rainforest climates with the blue fields to be on the east side of the island, starting at the northeast with Kukuihaele, a village above Waipi'o valley near the coast of Mt. Kohala. That diagram states a mean annual rainfall of 2,602 mm (104 in), at the top of the diagram. This still supports tropical rainforest in spite of the two short dry seasons there. The rainfall pattern there is similar to that shown on the climate diagram of the Hawai'i National Park Headquarters. It states 2,375 mm (95 in) mean annual rainfall and indicates a

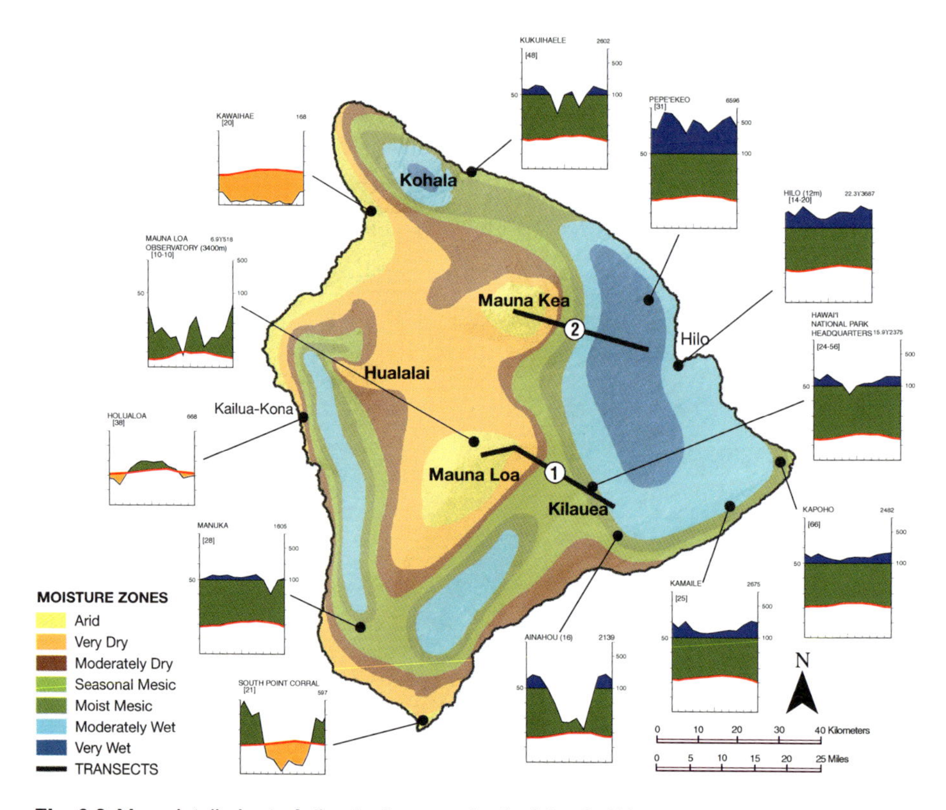

Fig. 3.2. More detailed set of climate diagrams for the island of Hawai'i superimposed on a map of moisture zones identified by Price et al. 2012.

short dry season (<100 mm, <4 in) in June. A more important difference is the mean annual temperature curve, which at the Park Headquarters is lower, recording a mean annual temperature of 15.9 °C. The climate diagram at ʻĀinahou in the National Park with 2,139 mm mean annual rainfall indicates a moderately dry seasonal environment, because of the prolonged dip below 100 mm/month during the summer months. ʻĀinahou is thus outside the rainforest territory.

Three other interesting climates deserve mentioning here. The climate at Kawaihae near the northwest tip of the island indicates a complete climatic desert like those found in the Sahara Desert. As seen on that diagram, the rainfall curve stays below the temperature curve throughout the year and the mean annual rainfall is a mere 168 mm. The other interesting climate diagram worth pointing out is at the Mauna Loa Observatory, which is famous for recording carbon dioxide amounts in the atmosphere. Here, the temperature curve stays very low year round with a mean annual value of 6.9 °C (44.4 °F). The diurnal temperature fluctuation exceeds the annual fluctuation and there is ground frost each night of the year. The month-to-month rainfall also fluctuates but it is always below 100 mm/month and averages only 518 mm/year (20.7 in). This environment indicates a tropical alpine desert. The third interesting climate is that of Kailua-Kona. In contrast to the other leeward climates it shows the reversal of the rainy season. In Kailua-Kona, the summer is moist and rainy while the winter is dry, often with drought conditions throughout these months.

Now that we have an idea of how the climate is distributed over the island, we will focus on two mountain transects. These topographic transects illustrate the kinds of ecosystems that co-occur with the native rainforest on the windward landscapes of the two highest Hawaiian mountains.

The Mauna Loa Transect (line 1 on the map, Fig. 3.2) and the Mauna Kea Transect (line 2 on the map) start at the top of each mountain and extend down into the rainforest. They show the Hawaiian rainforest in an altitudinal landscape perspective. Rainforest is only one among several other ecosystems on the windward slopes of these two huge (> 4,000 m, almost 14,000 ft) shield-shaped volcanoes.

The Mauna Loa Transect—going up the mountain

Well-preserved rainforest is restricted to the Kīlauea volcano at the Crater Rim Road (see Fig. 3.3). Segment 12 on the transect diagram refers to this area as depicted on Photo 3.1. Segment 11 relates to a more open grown rainforest with Uluhe fern in the undergrowth.

From here onwards on Highway 11 we are out of the rainforest climate. Going up the Mauna Loa Strip Road we come to the Bird Park (Kīpuka Puaulu) (Photo 3.2), and then to Kīpuka Ki (Photo 3.3), both are interesting forest and savanna types of vegetation, depicted on the Transect as segments 8 and 9. Here the mean annual rainfall is reduced to 1500 mm (60 in), and the climate is moderately seasonal with a slight summer-dry season. Both Kīpuka have deep soils from volcanic ash, which were identified as former ash dunes. The dominant trees here are Koa (*Acacia koa*) and Mānele (*Sapindus saponaria*).

Separating the two kīpuka is the 575-year old Keamoku lava flow. It was stocked until recently by a healthy ʻŌhiʻa lehua (*Metrosideros polymorpha*) forest with native shrub undergrowth of mostly Pūkiawe (*Leptecophylla tameiameiae*) and ʻAʻaliʻi (*Dodonaea viscosa*). Tree ferns and Uluhe ferns are absent here, clarifying that this was never a rainforest. In fact it is a forest much drier than the neighboring kīpuka forests in spite of both being in the same climate.

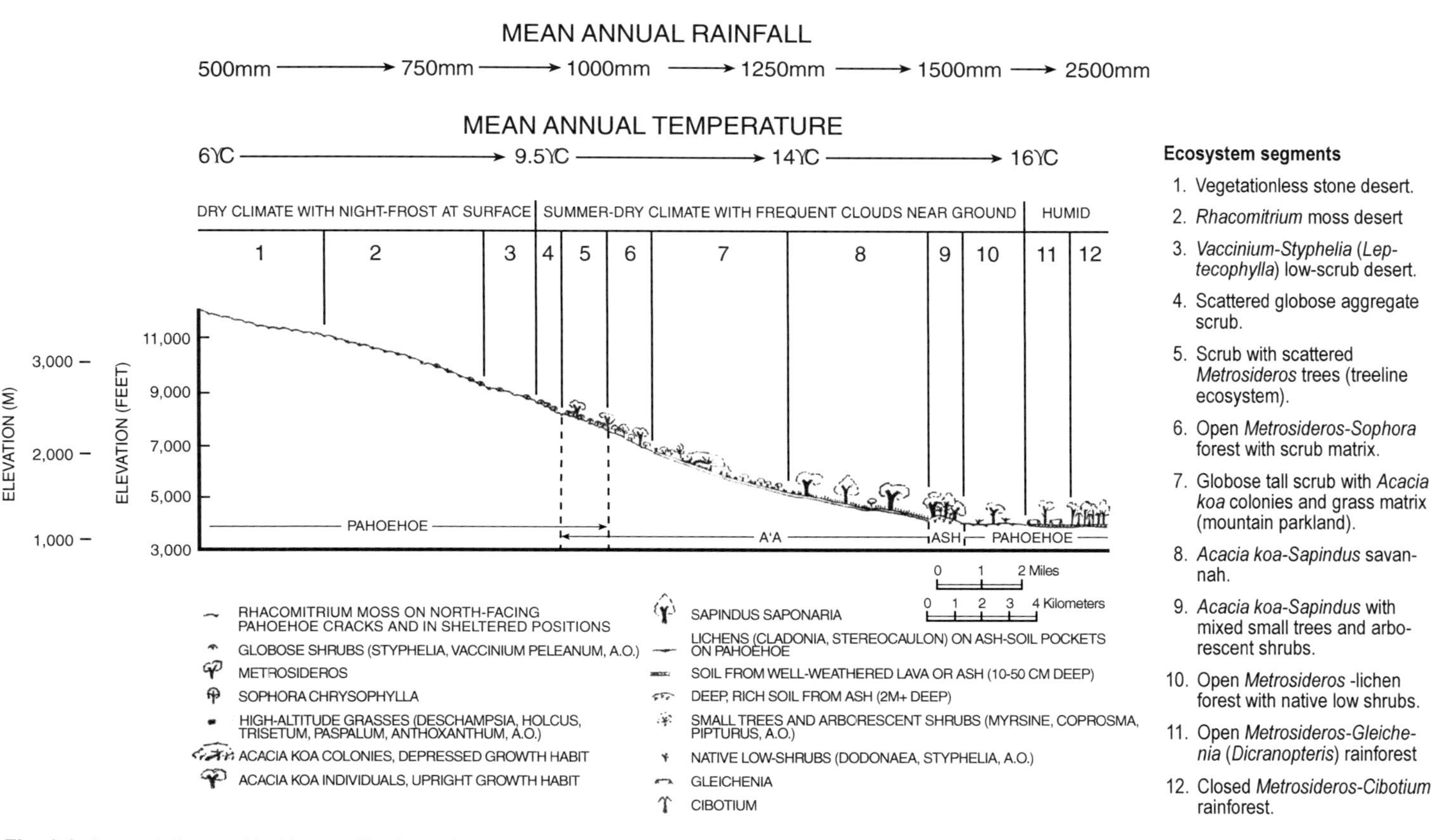

Fig. 3.3. A vegetation and habitat profile down the east slope of Mauna Loa volcano showing the 12 ecosystem segments that were studied during the Hawai'i IBP research program.

Photo 3.1. Rainforest near Thurston Lava Tube in Hawaiʻi Volcanoes National Park. This is part of the 200 year old mature forest along the Crater Rim Road and served as sample area for the Hawaiʻi IBP research in 1971–81. Photo taken in 1966.

Photo 3.2. Entrance to Kīpuka Puaulu (Bird Park) in Hawaiʻi Volcanoes National Park. Kīpuka Puaulu starts behind the ʻŌhiʻa dieback stand which is on the Keaumoku flow. Inside, the kīpuka is the same healthy forest as in Kīpuka Ki in Photo 3.3.

Photo 3.3. Young Koa trees growing along the side of the road in Kīpuka Kī, Hawaiʻi Volcanoes National Park. Photo taken in 1966.

The drier status of the ʻŌhiʻa forest on the Keamoku flow is due to its substrate, which is a well-to-excessively drained ʻaʻā flow. Within the last decade this forest has undergone canopy dieback. Most of the ʻŌhiʻa trees are now dead but still standing (Photo 3.4 and 3.5). The etiology of this forest dieback has not yet been investigated, but it may become if research is conducted in this area, similar to that described in Chapters 4–6, which deals with the investigation of a former rainforest dieback.

Going further up the transect, one comes through an ecosystem known as the mountain parkland (segment 7) (Photo 3.6). It is occupied by Koa colonies that expand via root suckers in a matrix of grassland. Fire was a natural disturbance here prior to arrival of the indigenous Hawaiians. Fire kills the Koa trees, but so far, Koa has always come back from cable-like roots of a few survivors. At about 2,000 m (6,560 ft) elevation near the end of the paved road comes another vegetation and sub-

Photo 3.4. Recent Ōhi'a canopy dieback on the Keamoku lava flow, which separates Kīpuka Puaulu (Bird Park) from Kīpuka Ki on the Mauna Loa Strip Road above the 1,220 m (4,000 ft) elevation marker in Hawai'i Volcanoes National Park.

Photo 3.5. Another view of the extensive 'Ōhi'a dieback on the Keamoku flow along the Mauna Loa Strip Road in Hawai'i Volcanoes National Park.

Photo 3.6. View of the mountain parkland ecosystem looking downslope from 2,000 m (6,700 ft) elevation along the Mauna Loa Strip Road in Hawaiʻi Volcanoes National Park. The dominating shrub is Pūkiawe (*Leptecophylla tameíameiae*). Grass patches are composed of native hair grass (*Deschampsia nubigena*) and introduced common velvet grass (*Holcus lanatus*). Koa trees now form large colonies in this community, from vegetative clones of trees that are connected by underground roots.

Photo 3.7. Open subalpine forest and shrubland with a broadly branched Ōhiʻa lehua tree. The person standing by the tree is Dr. Vladimir J. Krajina, July 1966. This photo corresponds to segment 6 on the Mauna Loa profile diagram (Fig. 3.3).

strate change to subalpine forest scrub on mostly pāhoehoe lava with almost no soil (Photo 3.7).

This shrubland with open grown 'Ōhi'a trees thins out more and more going upwards to the treeline ecosystem (segment 5). Here a few widely scattered 'Ōhi'a trees, from 3–5m tall with bushy crowns form thc upper tree line on Mauna Loa at 2,450 m (8,100 ft) (Photo 3.8). Further up shrubs become very low-growing and scattered to 3,050 m (10,000 ft) (Photo 3.9). Above that elevation the alpine stone desert begins with the moss *Rhacomitrium lanuginosum* growing in the windward crevasses of the older, buff colored pāhoehoe lava (Photo 3.10). The moss looks from a distance like small pockets of snow on account of their translucent (hyaline) leaf tips. Going still higher beyond 3,350 m (11,000 ft), the alpine stone desert (segment 1) presents an almost totally abiotic environment. On the trail leading to the old shelter near the Mauna Loa summit crater at 4,150 m (13,680 ft) (Photo 3.11), one can encounter a few horse droppings probably dating back to the early 1900s, when the summit shelter served for geological explorations. The horse droppings here are bleached by the sun but are still not decomposed.

THE IBP MAUNA LOA TRANSECT

The Mauna Loa Transect served as one of two major study sites for the International Biological Program in Hawai'i, known as the Hawai'i IBP. This multidisciplinary research program was conducted here from 1971–1981. The Mauna Loa Transect served for an integrated analysis of 14 organism groups (plants, birds, rodents, wood boring beetles, soil arthropods, soil microfungi and eight others) at 14 focal sites. These focal sites were distributed from the rainforest at Thurston Lava Tube (at 1,200 m, 3,900 ft) to the Pu'u 'Ula'ula (Red Hill at 3,050 m, 10,000 ft). (See map, Fig. 1.3.) The results were synthesized in a book titled *Island Ecosystems—Biological Organization in Selected Hawaiian Communities* (more in attached Bibliography).

Photo 3.8. This was one of the highest trees found in 1966 at 2,450 m (8,100 ft) elevation along the Mauna Loa trail in Hawaiʻi Volcanoes National Park. ʻŌhiʻa lehua is the only tree species found here. Such trees form the tree line community which has a scattered native shrub understory. The trees are very scattered at this elevation, approximately one tree/4,000 square meters (1 per acre), and range in size from 3–5 m (10–16.5 ft) tall.

Photo 3.9. Sparse alpine scrub at approximately 8,500 ft (2,575 m) along the Mauna Loa trail in Hawaiʻi Volcanoes National Park. This area corresponds to segment 3 on the IBP Mauna Loa Transect Profile, which extends from 8,500 ft (2,575 m) up to Puʻu ʻUlaʻula (Red Hill) cinder cone at 3,050 m (10,000 ft) elevation.

Photo 3.10. The moss *Rhacomitrium lanuginosum* grows in more protected windward crevasses of the alpine stone desert above 3,048 m (10,000 ft). The white translucent tips of this moss look like remnant patches of snow.

Photo 3.11. Moku'āweoweo crater dominates the summit of Mauna Loa volcano at 13,679 feet (4,172 m) elevation.

Note that the Hawaiian rainforest occupies only a small segment at the lower end of the Mauna Loa Transect. The reason for this is the limited extent of the rainforest climate in this area. It extends from the Kīlauea Rim forest north across Volcano Village and from here across the Saddle Road and further along the mid-slope area on Mauna Kea in a broad belt from sea level to 1,900 m elevation.

To include more of Hawaiʻi's rainforest in the Hawaiʻi IBP, the program was focused also on a rainforest study in an 80 hectare (200 acre) plot north of the National Park in the Kīlauea forest (owned by Kamehameha Schools, formerly named Bishop Estate). This forest is an "old-growth rainforest." Here Koa trees were dominating the canopy. ʻŌhiʻa lehua was second in the canopy. Many of the Koa trees were in a senescing life stage (see photo 1.20 on page 18). The dynamics of the Kīlauea rainforest was discussed in the book *Island Ecosystems* (see Mueller-Dombois et al. 1981).

The Mauna Kea Transect—coming down the mountain

This topographic ecosystem profile (Fig. 3.4) runs from the summit of Mauna Kea at 4,205 m (13,796 ft), north of the Saddle Road and from mid-slope down close to the Wailuku River to above the Hāmākua coast, where the sugar cane fields used to be 30 years ago.

Mauna Kea is a much older volcanic mountain that is currently considered dormant (Photo 3.12). During the last ice age, a little over 10,000 years ago, its summit area was under a glacier (Photo 3.13). Glacial striae (scratch marks) are still visible on some rock surfaces (Photo 3.14). The glacier retreated about 10,000 years ago. Despite the porous nature of the cinder substrate in much of the summit area, Lake Waiau at 3,970 m (13,020 ft) exists due to the locally low permeability of the substrate, likely very fine ash and clay (Photo 3.15 and 3.16). The

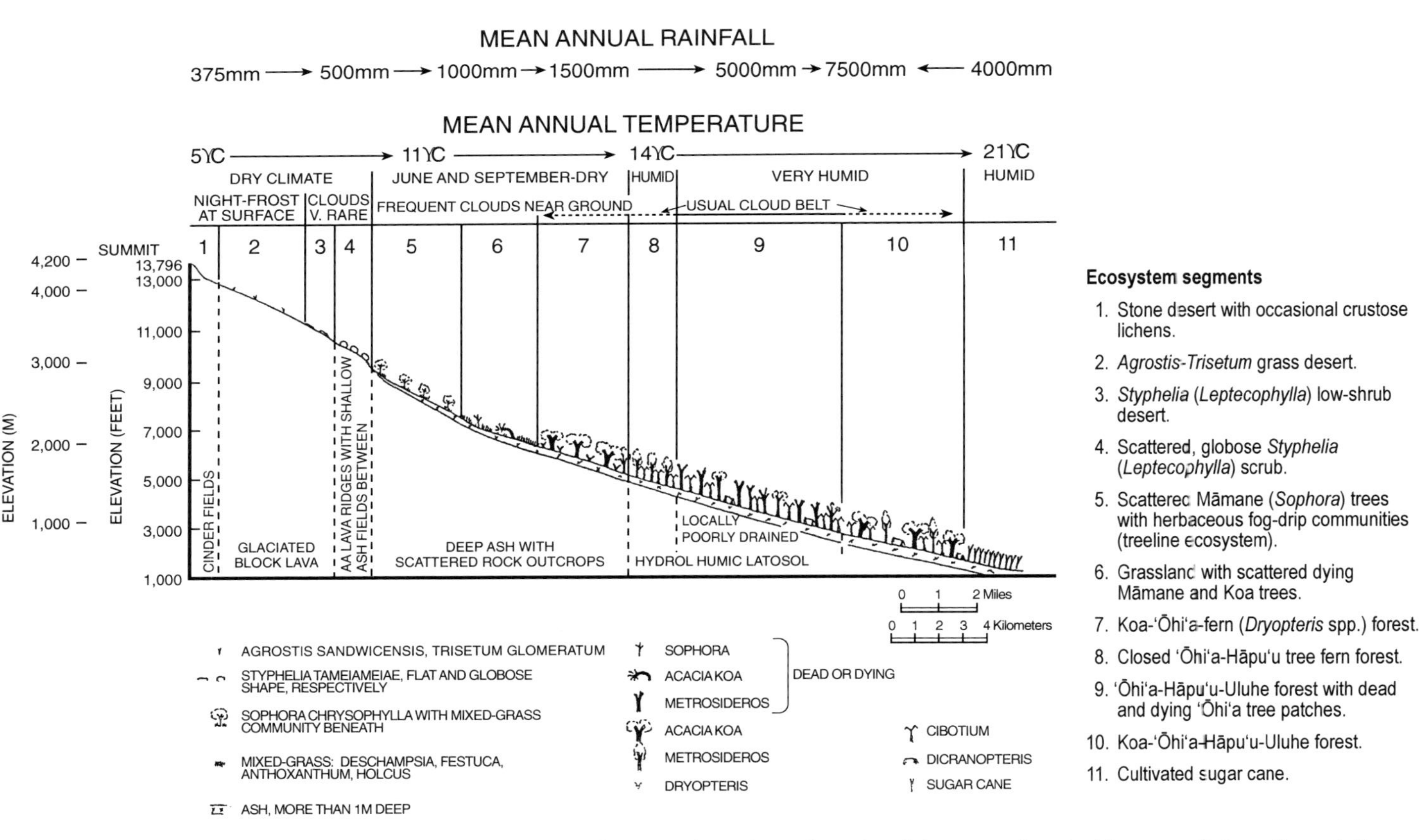

Fig. 3.4. A vegetation and habitat profile down the east slope of Mauna Kea volcano from the alpine stone desert at the summit down through the wet forest and dieback areas below 1,500 m (4,920 ft) elevation.

Photo 3.12. The summit of Mauna Kea is at the top of the Puʻu Wēkiu cinder cone at 4,205 m (13,796 ft) elevation.

Photo 3.13. View of the glaciated block lava desert below the summit cinder cone of Mauna Kea in 1966.

Photo 3.14. In the alpine stone desert, basalt rocks like this one show striations or grooves from past glacial activity in the summit area of Mauna Kea.

Photo 3.15. Lake Waiau, located at 13,020 feet (3,970 m) near the summit of Mauna Kea, is the only alpine lake in the Hawaiian archipelago and another residual feature of the last ice age, which ended about 10,000 years ago.

Photo 3.16. Dieter Mueller-Dombois measuring the surface water temperature of Lake Waiau in August 1966 as 15 °C (59 °F) while air temperature was 12 °C (54 °F). Photo courtesy of V. J. Krajina.

lake has decreased in volume in recent years, probably due to the combined effects of drought and increased temperature.

In its later phases of eruptive activity Mauna Kea was building cinder cones rather than extruding lava flows. Thus, the substrate from the summit down to the rainforest is a much improved medium for vegetation development as compared to hard-rock surfaces of Mauna Loa.

The next two profile segments (Fig. 3.4) are stone deserts of glaciated block lava with very sparse growth of some native high-altitude grasses (*Agrostis sandwicensis* and *Trisetum glomeratum*) (Photo 3.17).Then follows alpine scrub vegetation dominated by Pūkiawe (*Leptecophylla tameiameiae* formerly named *Styphelia tameiameiae*). Rarely one may find an ʻŌhelo ʻai (*Vaccinium reticulatum*) shrub clinging to a rock niche protected from browsing by the introduced feral sheep and mouflon (*Ovis* spp.) that still roam on Mauna Kea (Photo 3.18). The treeline ecosystem begins at 2805 m (9200 ft) with Māmane (*Sophora chrysophylla*) growing in parkland formation (segment 5 on profile diagram) (Photo 3.19 and 3.20). Here on Mauna Kea the tree line is more than 335 m (1,100 ft) higher than on Mauna Loa, where it is formed by 'Ōhi'a lehua alone. Most likely, 'Ōhi'a lehua was once also the first high-elevation tree on Mauna Kea. Later, Māmane became established, thereby replacing 'Ōhi'a in the course of primary succession.

Photo 3.17. Pili uka (*Trisetum glomeratum*), a native high altitude grass growing sparsely in the vicinity of Lake Waiau near the summit of Mauna Kea.

Photo 3.18. An old individual of the native shrub ʻŌhelo ʻai (*Vaccinium reticulatum*) growing in a rock crevice protected from browsing by feral sheep. This species is a member of the alpine scrub community that grows above tree line on the upper elevation slopes of Mauna Kea.

Photo 3.19. View down to the tree line ecosystem on Mauna Kea which, in this area, is dominated by scattered Māmane (*Sophora chrysophylla*) trees. The tree line ecosystem on Mauna Kea extends to 9,200 ft (2,805 m) elevation, which is approximately 1,100 ft (335 m) higher in elevation than the tree line on Mauna Loa, seen in the background. In the foreground of the picture are two native Pūkiawe shrubs (*Leptecophylla tameiameiae*). Pūkiawe dominates the alpine scrub ecosystem that merges into the alpine stone desert above 3,353 m (11,000 ft) elevation.The spiny foliage of Pūkiawe protects the bushes from overgrazing by feral sheep and the introduced mouflon sheep.

Photo 3.20. View of the tree line ecosystem on Mauna Kea. The light patches under the Māmane trees are fog-drip communities composed of introduced European grass species that thrive under the tree canopy that intercepts moisture from clouds that come close to the ground as fog at this elevation.

As we come further down the mountain, tree growth becomes very sparse (segment 6). There are a few senescing and dead trees, mostly of Māmane (Photo 3.21), but also of Koa still standing in a matrix of pasture grass (Photo 3.22). The dying of trees there is easy to explain. Any reproduction from seedlings or suckers has no chance of surviving, because of the continuing grazing pressure by introduced cattle and sheep. It is rather surprising that there are still some surviving trees. If management of this area would include fencing out the grazers, forest groves could likely become reestablished without much further assistance. Profile segment 7 represents a Koa savanna more or less well protected from grazing by a fence. It grades into mixed Koa-ʻŌhiʻa old-growth rainforest indicated as segment 8 on the Mauna Kea profile diagram. The latter two segments refer to the Hakalau forest, managed for restoration and preservation by the US Fish and Wildlife Service.

Photo 3.21. A senescing Māmane tree, a remnant of the former subalpine Māmane-Koa forest on the mid-elevation slopes of Mauna Kea. This formerly closed canopy forest has been opened up over the years by extensive grazing by cattle and sheep, and occasional fires.

Photo 3.22. A view of what is now subalpine pastureland on Mauna Kea above 1,981 m (6,500 ft) elevation. This landscape is above the general cloud and inversion layer. The trees found here are senescent and dead trees of Koa, Māmane, and a few Naio (*Myoporum sandwicense*) trees, which are the remnants of a former subalpine forest that was opened as a result of extensive grazing by cattle and sheep and occasional fires.

From here we come into segment 9, the wetter rainforest where Koa is rarely present. Instead, ʻŌhiʻa dominates the canopy outright. Groups of dead and dying ʻŌhiʻa trees were detected here when the transect was established in 1966 (Photo 3.23, 3.24, and 3.25). This ʻŌhiʻa dieback forest, which is comprised of senescing stands of trees, is clearly noted on the profile diagram in Fig. 3.4.

Another mixed ʻŌhiʻa-Koa forest was encountered further down, designated as segment 10. It appeared to be similar to that at higher elevation in segment 7, but the lower forest was more disturbed. The trees were more irregularly spaced, and larger openings were filled with thickets of Uluhe fern. The lower forest boundary was fringed by a row of planted Swamp mahogany trees (*Eucalyptus robusta*, introduced from Australia) and cut off by sugar cane fields.

Photo 3.26, taken in October 1984 by Kim Bridges on a cloudless day, gives a broad landscape perspective of the east

Photo 3.23. View of dying ʻŌhiʻa rainforest in 1966 around 1,200 m (4,000 ft) on a 3,000 year old lava flow near the Saddle Road on the island of Hawaiʻi. Note the healthy undergrowth of tree ferns and the large-diameter tree in the center of the photo.

THE IBP STABILITY VERSUS FRAGILITY DEBATE REGARDING ISLAND ECOSYSTEMS

The Hawaiʻi/IBP members (about 25 senior scientists and 25 graduate students) had a conceptual discussion following their initial field trip in August 1970. The discussion centered on evolution and the stability/fragility concept of island ecosystems. The prevailing view was that island ecosystems are fragile relative to continental ecosystems. The greater stability of the latter was to be found in their larger species diversity, which also would be reflected in a greater functional diversity and thus stability. Investigating the ecological and evolutionary status of endemic species was considered a critical research component. Based on this concept, it was argued, that the initial focus should be on the biological organization of intact island ecosystems. In other words, problem ecosystems such as the forest with dieback, should at this time be left to the plant pathologists and forest pest researchers. Instead the IBP focused on an altitudinal sequence of 14 focal sites along the Mauna Loa Transect inside the National Park and on a healthy rainforest in the Kīlauea forest reserve near the Transect (see Fig. 1.3, p. 22). This Kīlauea rainforest had a senescing cohort of Koa trees mixed with healthy ʻŌhiʻa trees in the canopy. Here the Koa trees were intended for harvesting by the land owner, the Bishop Estate. After some discussion to postpone the logging of Koa, the Kīlauea forest was offered for study to the IBP team.

Photo 3.24. An aerial view of 'Ōhi'a dieback forest on the island of Hawai'i showing the cohort stand structure of the canopy trees. Note the healthy undergrowth of mostly Hāpu'u tree ferns. Photo taken by J. Jacobi, 1976.

Photo 3.25. 'Ōhi'a dieback forest in a study plot located along the Tree Planting Road on the island of Hawai'i. Photo by J. Jacobi.

SENESCING STANDS can also be considered old-growth forests, but they form a uniformly structured cohort of 'Ōhi'a trees that may have suffered from habitat constraints and/or intra-specific competition, both of which can cause premature senescence, and, ultimately, death. Habitat constraints include nutrient and soil water regime imbalances that occur naturally with soil- and landscape aging. Large-diameter 'Ōhi'a trees, likely twice as old, are found surviving in the wider area of senescing forests. They are surviving in the same habitats with the senescing cohorts but merely as sporadic remnants of an earlier closed forest. Photos 3.23 and 3.24 illustrate this point. In senescing cohort forests, canopy gaps are created by trees that are losing or have lost their crown foliage but remain standing as snags. These typically result in large (or small but repetitive) gaps over wider areas depending on the cohort structure. In old-growth forests, gaps are typically outlined by fallen trees. This small area succession or turn-over pattern is often referred to as "gap dynamics." In senescing cohorts, we refer to the turnover pattern as "dieback dynamics."

Photo 3.26. View of the entire windward slope of the huge Mauna Kea shield volcano on a cloudless day. Shown are the major vegetation zones ranging from sugar cane fields above the ocean, the broad rainforest belt in the middle, the light green colored pastureland above the forest, which then grades into the Māmane forest and the darker alpine shrubland communities, followed upwards by the stone and cinder desert near the summit of the volcano which is over 4,000 m (13,123 ft) tall. Photo courtesy of K. W. Bridges, 1984.

slope of Mauna Kea where the transect was done. It starts above the ocean with the sugarcane fields. Upslope extends the rainforest belt. Above the rainforest belt is a pale green pasture belt topped by a darker green-gray alpine scrub belt. The pale orange summit area marks the alpine stone and cinder desert. The rainforest belt is normally covered by a layer of clouds. Its upper limit coincides with the inversion layer near 1,900 m elevation. This layer is about halfway up this huge shield-shaped mountain. The 'Ōhi'a dieback became a research priority in the 1970s through 1980s as will be explained in the following chapters.

Suggested Readings

Juvik, S. P. & Juvik, J. O. (1998). *Atlas of Hawai'i*. 3rd Edition. Honolulu: University of Hawai'i Press. 333 p.

Kellner, J. R., Asner, G. P., Vitousek, P. M., Tweiten, M. A., Hotchkiss, S., Chadwick, O. A. (2011). Dependence of forest structure and dynamics on substrate age and ecosystem development. *Ecosystems* 14(7): 1156–1167.

Kitayama, K. & Mueller-Dombois, D. (1995). Vegetation changes along gradients of long term soil development in the Hawaiian montane rain forest zone. *Vegetatio* 120: 1–20.

Krajina, V. J. (1963). Biogeoclimatic zones on the Hawaiian islands. *Newsletter of the Hawaiian Botanical Society*. II(7): 93–98.

Mueller-Dombois, D., Bridges, K. W. & Carson, H. L. (eds.) (1981). *Island Ecosystems: Biological Organization in Selected Hawaiian Communities*. Stroudsburg, PA & Woods Hole, MA: Hutchinson Ross Publishing Company, US/IBP Synthesis Series 15. 583 p.

Mueller-Dombois, D. & Fosberg, F. R. (1998). *Vegetation of the Tropical Pacific Islands*. New York: Springer-Verlag New York, Inc. 774 p.

Mueller-Dombois, D. & Krajina, V. J. (1968). Comparison of east-flank vegetations on Mauna Loa and Mauna Kea, Hawaii. In *Proc. Symp. Recent Advances in Trop. Ecol.*, ed. by R. Misra and B. Gopal. Vol. II. Varanasi: International Society for Tropical Ecology. pp. 508–520.

Mueller-Dombois, D. & Lamoureux, C. H. (1967). Soil-vegetation relationships in Hawaiian kipukas. *Pacific Science* 21: 286–299.

Price, J. P., Jacobi, J. D., Gon III, S. M., Matsuwaki, D., Mehrhoff, L., Wagner, W. L., Lucas, M. & Rowe, B. (2012). Mapping plant species ranges in the Hawaiian Islands-Developing a methodology and associated GIS layers. US Geological Survey, Reston, VA. http://pubs.usgs.gov/of/2012/1192/of2012-1192_text.pdf

Ripperton, J. C. & Hosaka, E. Y. (1942). Vegetation zones of Hawaii. *Hawaii Ag. Exp. Sta. Bull.* 89: 1–60.

Vitousek, P., Asner, G. P., Chadwick, O. A. & Hotchkiss, S. (2009). Landscape-level variation in forest structure and biogeochemistry across a substrate age gradient in Hawaii. *Ecology* 90: 3074–3086.

Chapter 4

Collapse in the Hawaiian Rainforest

The ʻŌhiʻa canopy dieback, noticed first in 1964 along the Mauna Kea Transect, received a lot of attention from forest researchers in the early 1970s and throughout the following decade. The initiating event for this attention was the first IBP field trip.

This field trip was conducted along the Mauna Kea Transect for the purpose of establishing research sites for the Hawaiʻi IBP. A group of senior researchers, as well as some graduate stu-

Photo 4.1. View of the zone of ʻŌhiʻa forest dieback in the mid-1970s stretching from the middle slopes of Mauna Kea in the foreground across to the northeastern slope of Mauna Loa. Recent lava flows, less than 200 years old, cut through the forest in the middle of the picture. Photo by J. Jacobi.

dents took part. A vivid discussion arose as the group entered transect segment 9 and noted the canopy dieback there. The majority of fieldtrip participants considered this to be a new "killer disease" that had infected the 'Ōhi'a trees. A minority considered a "natural cause" as published in 1968 (by Mueller-Dombois & Krajina, see Appendix C).

A quote from that paper will clarify this point: "Here the rainfall is very high (5,000 mm = 200 in), fog is almost constantly present, and soil drainage is impeded. It is the area on the Mauna Kea slope where numerous stream beds begin to form, but where they are not yet deeply cut into the substrate." In other words the canopy dieback was considered to have a "natural cause," i.e., flooding of the root system in an area with almost constant fog drip where soil drainage was stagnating due to as yet incomplete stream development.

The field discussion ended with a split. The forest pathologists opted to write a new proposal to obtain research funding for the "killer disease hypothesis." The IBP group opted to concentrate instead on healthy intact ecosystems. There was a general assumption in the IBP group that island ecosystems are fragile and that it was better to focus first on ecosystems without such problems as indicated by the canopy dieback.

Research to detect the killer agent

Early results

The search to locate disease or insect pest agents responsible for the dieback began with pathologists taking soil and root samples from around dying trees. Entomologists also searched for pest insects. Another approach was an air photo analysis of the windward rainforest territory by foresters of the US and Hawai'i State government.

Phytophthora cinnamomi, a root pathogen that was known to have caused jarrah *(Eucalyptus marginata)* forest dieback in

Western Australia, was found early in the investigation. Air photo sets covering the area from the Hāmākua coast upslope of Laupahoehoe at the northern end to the rainforest in Hawai'i Volcanoes National Park at the southern end, were available from 1954 and 1965.

With new government funding for investigating the killer disease problem, a third air photo set was obtained in 1972. A phenomenal increase of canopy dieback or forest decline (as the dieback was also called) was discovered from these air photos. In an 80,000 hectare (ha) sample from north to south, the photo analysis revealed a progression of severe decline from 120 ha in 1954 to 16,000 ha in 1965, and 34,500 ha in 1972. The investigators of these early results concluded in 1975 that "if the decline continues at the present rate, remaining 'Ōhi'a forest in the study area will be virtually eliminated in 15–25 years" (E. Q. P. Petteys, R. E. Burgan & R. E. Nelson 1975. See Appendix C, p. 239–241).

It seemed that a rapid spread of the same killer fungus responsible for the jarrah dieback in Western Australia was suddenly at work in our native 'Ōhi'a rainforest; certainly a frightening prospect (Photo 4.2).

Follow-up results

The 'Ōhi'a borer *(Plagithmysus bilineatus)* (Photo 4.3), an endemic "longhorn" cerambycid beetle, was also found to be frequent in dieback stands and thus considered to be responsible in part for the epidemic decline of 'Ōhi'a forest. But the 'Ōhi'a borer occurred also in healthy stands along the Mauna Loa IBP Transect. After sifting through a number of possible biotic agents, including bacteria and nematodes as well as a variety of wood-decay fungi, the root pathogen *Phytophthora cinnamomi* and the endemic wood-boring beetle *Plagithmysus bilineatus* remained as the most suspicious culprits.

Photo 4.2. Small patches of "hotspot" dieback in the dense ʻŌhiʻa forest near Puʻu Makaʻala off Stainback Road on the island of Hawaiʻi. Initially these "hotspots" were thought to be inoculation points from which fungal pathogens, such as *Phytophthora cinnamomi*, might spread. Photo by J. Jacobi.

Photo 4.3. The endemic two-lined Ōhiʻa borer *Plagithmysus bilineatus*, once suspected to be a primary agent causing the Ōhiʻa dieback. However, this beetle lays its eggs in dying Ōhiʻa trees and, therefore, has merely a hastening role in the death of the trees. Photo by Karl Magnacca, University of Hawaiʻi at Hilo.

THE MEDIA TOOK HOLD OF THE PROBLEM

First, the *Honolulu Star-Bulletin* came out with the headline:

Tree Killer Perils Isle Koa and ʻŌhiʻa
September 29, 1970

> "Disease is sweeping the Big Island's native ʻŌhiʻa and koa forests, killing thousands of acres of trees and resulting in immeasurable loss...."

Other headlines:

Death of ʻŌhiʻa Forests a Frightening Prospect
Honolulu Star-Bulletin, May 17, 1975

> "State and US foresters are frightened by the rapidly spreading death of native ʻŌhiʻa forests on the Big Island ...."

Hawaiian Rainforests are in Trouble
Christian Science Monitor, December 1975

> Plant pathologists call it an "epidemic state of decline."

ʻŌhiʻa Trees are Dying
Honolulu Star-Bulletin, March 7, 1975

> "ʻŌhiʻa LEHUA, the most common native tree in the Hawaiian mountains, is the pioneer tree on new lava flows. It is thus associated with life.
>
> But there is another aspect to ʻŌhiʻa. In ancient Hawaiʻi the first man killed in battle was called the LEHUA. In this fashion it was associated with death...."

For more details, see Appendix D, p. 265.

The entomologists working on the evolutionary significance of the cerambycid beetles (J. Lindsay Gressitt, Cliff Davis and G. Allan Samuelson) remarked that the ʻŌhiʻa dieback helped to get better access to this otherwise rare and evolutionarily significant member of the wood-boring cerambycid beetles. Most of the endemic cerambycid beetles are host-tree specific and considered to be normal, but rare, members in native Hawaiian ecosystems (see IBP Synthesis Volume page 135).

Further research on the distribution and prevalence of *Phytophthora cinnamomi* revealed that this root fungus is not always associated with dying ʻŌhiʻa trees or dieback stands, but that it also occurs in perfectly healthy stands. It was found to be more associated with soil moisture regimes, not with the dry and wet types, but rather with those intermediate.

Two pathologists (Wen H. Ko and J.T. Kliejunas), sceptical of the killer role of *Phytophthora cinnamomi*, fertilized a group of dying trees on a well-drained young pāhoehoe flow with a normal dose of N-P-K (nitrogen-phosphorus-potassium). They found these trees to resprout new leaves after they had lost most of their crown foliage. An adjacent group of dying trees was treated with fungicide, and there was no foliage recovery. These authors concluded that trees were dying from nutrient starvation of nitrogen and phosphorus.

Moreover, a dissertation on the Hawaiian strain of this root pathogen revealed that it was not an obligatory parasite. Instead it was surviving well in dead organic matter. This meant that this strain of *Phytophthora cinnamomi* was rather harmless in the Hawaiian rainforest.

The final conclusion of a decade of intensive search for the killer agent was that neither of the two suspected biotic agents was the trigger of the epidemic ʻŌhiʻa decline, and that the two agents could be involved only after the trees would succumb to an "**unknown stress factor**" (R. P. Papp, J. T. Kliejunas, R. S. Smith, Jr. & R. F. Scharpf 1979).

Research to detect natural dieback causes as an alternative to the disease hypothesis

When funding for the Hawaiʻi/IBP came to an end in 1974, a new research proposal was written to tackle the ʻŌhiʻa dieback problem from a different perspective. The proposal was submitted to the National Park Service with the research hypothesis "that the 'Ōhi'a dieback is a normal phenomenon, a developmental stage in primary succession of an isolated rainforest ecosystem." In short, this was called "The Succession Hypothesis" as an alternative to "The Disease Hypothesis."

Significant findings

A habitat classification revealed that there were five different types of stand-level dieback. We named them Wetland dieback, Dryland dieback, 'Ōhi'a displacement dieback, Gap-formation dieback, and Bog-formation dieback.

Wetland dieback was the most common form of dieback (Photo 4.4, 4.5, and 4.6). This dieback type was restricted mostly to pāhoehoe lava substrates with pockets of mineral soil but a continuous cover of dead organic matter overlying the lava rock substrate. Water often stood above the rock surface during rainy periods. During prolonged dry periods the water disappeared from the surface but it was still present among the rock fissures; the organic overlay never dried up completely (Photo 4.7 to 4.10).

Photo 4.4. Wetland dieback as seen from the Saddle Road in November 1966 at about 1,150 m (3,500 ft elevation).

Photo 4.5. Wetland dieback as seen in 1977 along the Wailuku River road which runs north from the Saddle Road at approximately 1,150 m (3,500 ft elevation). Photo by J. Jacobi.

Photo 4.6. A view of 'Ōhi'a forest along the Wailuku River road in 2005. Notice that the dead trees seen in Photo 4.5 have mostly disappeared and have been replaced by young 'Ōhi'a that have grown since the tree canopy was opened by dieback in the early 1970s.

Photo 4.7. Jim Jacobi collecting data in wetland dieback plot #1 located along the Wailuku River road in January 1976.

Photo 4.8. Richard Becker and Ranjit Cooray standing in a wetland dieback site along Stainback Road showing advanced sapling regeneration in January 1974.

Photo 4.9 (left). Richard Becker in a wetland dieback site lifting a small Ōhiʻa sapling from the soggy ground exposing its flat, diminutive root system.

Photo 4.10. Wetland dieback on the windward side of the island of Hawaiʻi as seen from the air, 1984. Photo by J. Jacobi.

SUCCESSION is a fundamental ecological concept, which refers to the change in the biological composition of species and the changes in soil and environmental factors over time in the same area. An important ingredient is the "turnover" of individuals, populations and species, a process that involves death of individuals or groups and also rebirth. In ecology, space is often substituted for time. This means that instead of long-term studies of changes in the same area, changes are observed on differently aged soil substrates that are considered to represent a successional sequence, when in fact they are a time-sequence in space, i.e., a chronosequence. However, if this chronosequence is in the same area and moisture zone, one can consider a chronosequence to represent a long-term succession.

A HYPOTHESIS is a means to focus an investigation on a specific subject. As such it prescribes the methods to be used for the investigation. Simply put, a hypothesis is an "educated guess" in form of an assertive statement based on a research question. With strong conviction, a hypothesis can be in form of a "prediction." But it requires to be tested in any valid scientific investigation.

TO TEST THE SUCCESSION HYPOTHESIS a representative number of sample plots needed to be established throughout the dieback territory in dieback and healthy forest stands from volcanically young to old substrates. The focus was on complete soil, substrate, and vegetation analyses. This included two soil pits to be dug in each plot, a detailed floristic and tree structural analysis, a classification of the health status of individual trees, including seedling and sapling counts, mapping the 'Ōhi'a rainforest vegetation by structural criteria across the same area, and determining the percent cover of defoliated tree crowns by visual observations from the ground as well as by remote sensing techniques.

The results were published and also presented in a joint 'Ōhi'a DECLINE SEMINAR October 25,1977 that included forest ecologists, forest pathologists, and insect pest researchers (Mueller-Dombois et al. 1980).

Dryland dieback was less common, but also significant, particularly in view of the fact that this was not seen along the Mauna Kea Transect originally (Photo 4.11 and 4.12). Thus, the discovery of canopy dieback without flooding of the root system was a surprise, and gave temporary suspicion again that a biotic agent may be involved. Dryland dieback was found mostly on pyroclastic substrates, on 'a'ā lava and/or cinder deposits, both relatively young, well-drained substrates.

Photo 4.11 (left). Dryland dieback forest along Highway 11 near Volcano Village in 1976.

Photo 4.12. Dryland dieback along Highway 11 near Volcano village in 1989.

ʻŌhiʻa displacement dieback was first seen in the "old-growth forests" of the Ōlaʻa Tract belonging to Hawaiʻi Volcanoes National Park (see Figure 1.3) (Photo 4.13). Here, soil is rather nutrient rich (**eutrophic**) from volcanic ash and relatively young, perhaps 1,000-3,000 years old. It is moderately well drained and seldom flooded. The displacement relates to the fact that after ʻŌhiʻa dieback the undergrowth of Hāpuʻu tree ferns form the canopy, which may even increase its cover. ʻŌhiʻa seedlings, which need high levels of light to grow, stagnate under this dense shade and succumb also to physical damage from fallen tree fern fronds (Photo 4.14).

We conducted an experiment in which the tree fern canopy was partially removed and found that many shade-born ʻŌhiʻa seedlings died upon the sudden light influx, but others responded immediately with height growth. New light-born seedlings came up in large numbers and they grew faster than the shade-born individuals. The ʻŌhiʻa seedling displacement is never complete, even with a dense overstory of Hāpuʻu tree ferns, because the fern trunks and trunk tops serve as germination sites for ʻŌhiʻa seedlings. Thus, a new ʻŌhiʻa forest will eventually become rebuilt after the old trees die and fall down.

In contrast to the ʻŌhiʻa displacement dieback, both the dryland and wetland dieback types gave early indications that they could be regarded as ʻŌhiʻa "**replacement dieback**" types from their early changes in structure. In many plots, we observed under the dying canopy trees ʻŌhiʻa seedlings and saplings.

A new research grant from the National Science Foundation received in 1979 allowed for helicopter time and thus exploration of the more inaccessible dieback territory north of the Wailuku Stream on the older wet slope of Mauna Kea. Some of us made two helicopter trips with 2–3 nights camping in the more open bog areas. Here we got a close look at two more types of dieback.

Photo 4.13. 'Ōhi'a displacement dieback near the Volcano Agricultural Experiment Station in 1980. Notice the salt-and-pepper pattern of the dying trees. The substrate here is in the eutrophic phase of soil development.

Photo 4.14. Ground view of Ōhi'a displacement dieback from inside the Volcano Agricultural Experiment Station showing the vigorous Hāpu'u tree fern layer exposed under scattered dying 'Ōhi'a trees.

Gap-formation dieback was characterized from the air as small groups of 'Ōhi'a trees fallen or standing dead in a matrix of healthy forest (Photo 4.15). On the ground we found tree group dieback among healthy stands on moderately to poorly drained knolls and flats raised above the boggy terrain on the Mauna Kea slope in the 600–1,000 m (2,000–3,000 ft) elevation range above the Hāmākua Coast. Thus, in contrast to the often large (landscape size) dieback patterns on the wet windward slope of Mauna Loa on both sides of the Saddle Road, dieback on the older soils of Mauna Kea, north of the Wailuku stream, formed a much smaller mosaic pattern; another "salt-and-pepper" type as seen from the air.

These landform types (knolls, ridges, and flats) stood up several to tens of meters (~10–60 ft) above the boggy and permanently wet and soggy soil terrain (Photo 4.16). This kind of

Photo 4.15. Gap-formation dieback in 1974 on the windward, east slope of Mauna Kea. Here gaps in the forest are often filled with Uluhe fern. Compare to Photo 4.13, where Hāpu'u tree ferns dominate the understory in the displacement dieback.

gap-formation dieback was found to be prevalent also in sections on the Kohala mountain area. On Mauna Kea it occurred on the older, deeply weathered (**oligotrophic**) soils (Photo 4.17 and 4.18). These soils were nutrient impoverished, had lost most of their calcium and magnesium, and phosphorus was tied up in unavailable form. Aluminum and manganese were found to be in near-toxic proportions. Many 'Ōhi'a tree groups were dying and forming canopy gaps in the knoll-and-depression topography. We were surprised to find Koa trees also dying here in this very wet (5,000–7,000 mm/year rainfall area, i.e., 200–280 in/year).

The whole area seemed to be in a dynamic stage of stand-level dieback associated with geomorphological change. It seemed like a grand erosion pattern, the knolls, ridges and flats being remnant surfaces of the volcanic shield. The depressions

Photo 4.16. Annette Mueller-Dombois, Nadarajah Balakrishnan, and Jim Jacobi digging soil pits in January 1976 in a treeless sedge (*Carex* spp.) swamp near the Wailuku River. Gap-formation dieback can be seen starting in senescing 'Ōhi'a canopy cohort on the ash dune in the background.

Photo 4.17. Gap-formation dieback north of the Wailuku River at 1,200m (4,000 ft) elevation.

Photo 4.18. Joan Canfield in August 1980 counting seedlings in an area with abundant ʻŌhiʻa regeneration within a canopy gap in the same area as Photo 4.21 on the eastern slope of Mauna Kea.

and flats below and surrounding the knolls appeared to have lost segments of the original surface shield. One of our field research associates, Grant Gerrish, nicknamed this terrain combined with the next type of dieback "the ʻŌhiʻa graveyard area" (Photo 4.19, 4.20, and 4.21).

This nickname explained the drama that we witnessed on the ground during our helicopter explorations, but it missed the truth about what was going on. ʻŌhiʻa lehua trees were dying, many had fallen and turned into decaying wood and woody peat, but new individuals were being reborn as seedlings on decaying logs and adapting with stunted growth. By further reanalyzing the Maui forest trouble (see below), we learned what really was happening there.

Photo 4.19. Some larger canopy gaps in the regression phase become overwhelmed by the native Uluhe fern. Grant Gerrish is almost hidden by the deep mat-forming and tree climbing fern.

Photo 4.20. Joan Canfield and Grant Gerrish collecting data in a vegetation plot within a large canopy gap that has a large number of seedlings, indicating gap replacement with ʻŌhiʻa.

Photo 4.21. Gap-formation dieback with rainbow over the forest south of our camp on the evening of July 16, 1980.

Bog-formation dieback. Below the ridges, knolls, and moderately drained flats were larger areas of soggy flats. Here water was at the surface moving slowly over clay banks vegetated with broomsedge grass (*Andropogon virginicus*). The mineral soil on other low, typically much larger flats, was covered with woody peat tangled with and overlain by a matrix of Uluhe fern (*Dicranopteris linearis*). Here were many stunted (small 1–3m tall) 'Ōhi'a trees, also Koa branches, and tree fern trunks (Photos 4.22 through 4.25). Some taller and earlier fallen 'Ōhi'a trees had their former branches standing up as new trees, and the lower branches stuck in the peaty soil performing as roots. This situation we called bog-formation dieback (Photos 4.26 through 4.32).

The US Forest Service Team following independently with the same ecological approach, came up with essentially the same dieback types in their final report (Hodges et al. 1986).

Photo 4.22. Bog-formation dieback area as seen from air. Note the two clay bogs in the center of the photo (light brown areas), the Uluhe fern (light green areas) forming organic bogs that include fallen trees that died during an earlier 'Ōhi'a canopy dieback event. Photo taken in 2005 by Rick Warshauer.

Photo 4.23. Our field camp in a canopy gap near the bog-formation dieback area in the very wet mid-elevation slope of windward Mauna Kea, July 1980.

Photo 4.24. Lani Stemmermann examining a patch of Uluhe fern within a bog surrounded by dieback forest. These bogs drain into larger streams further downslope.

Photo 4.25. Bog-formation dieback with fallen 'Ōhi'a trees that survive by turning upright branches into clonal trees and downward branches developing roots.

Photo 4.26. Bog-formation dieback in the central Hāmākua forest on the eastern slope of Mauna Kea. This is an example of stand-reduction dieback in which the larger dead and dying 'Ōhi'a trees are replaced by smaller statured trees that can better tolerate the wet boggy conditions in the substrate.

Photo 4.27 (above). Ranjit Cooray sitting in an open patch within a stand of bog-formation dieback. Many of the branches on the fallen ʻŌhiʻa tree in the background have turned into vegetative clones of the original tree and are sending roots down into the ground.

Photo 4.28. Grant Gerrish standing in what he called the ʻŌhiʻa forest graveyard. He was correct in that this area appeared to be an incipient stream with complete die-off of the trees.

Photo 4.29 (left). Moderately drained bauxitic soil in a gap-formation dieback area on the east slope of Mauna Kea. A partial iron-aluminum hardpan is indicated by the rusty color in lower part of profile.

Photo 4.30. A clay bog soil which shows as a very dark color in the upper 30 cm (1 foot) due to enrichment of this layer by organic colloids. To the right of the pit are plastic bags containing soil samples that were collected for nutrient analysis back in the laboratory.

Photo 4.31. Another example of an organically enriched clay bog soil layer overlying red bauxitic soil with introduced rushes and sedges (*Juncus* and *Cyperus*) growing on the surface.

Photo 4.32. Study pit dug into an organic bog soil built up on top of former forest soil on the eastern slope of Mauna Kea in August 1980. Stagnant water seeped into the soil pit as soon as it was dug.

The Maui forest trouble

Early in the last century, 'Ōhi'a forest dieback was detected on East Maui, on the lower windward slope of Haleakalā mountain (Photo 4.33). Sugar cane plantation owners and managers were gravely concerned about loss of watershed values since the East Maui Irrigation System had just been installed. Was that dieback initiated by opening the closed forest allowing for new storm channels to affect the forest? That was one of the first questions.

When researchers began to investigate the Maui dieback, it had already affected an area from Kailua to Nāhiku over a stretch of 35 km (22 miles) above the winding East Maui Road to Hāna. The forest seemed to be dying in an elevation belt from

Photo 4.33. View of a former 'Ōhi'a dieback area on the windward slope of Haleakalā volcano on the island of Maui. Kanehiro Kitajama is standing among bushy 'Ōhi'a trees with Uluhe fern in the foreground in this photo taken in June 1988. Many more shrub-like, stunted 'Ōhi'a trees can be seen in the background with remnants of former tall 'Ōhi'a trees that died in the early 1900s. The trees at left and right of the photo are invading Paperbark trees (*Melaleuca quinquenervia*).

300–600 m (1,000–2,000 feet). Not only ʻŌhiʻa trees were dying, also the Hāpuʻu tree ferns and associated native trees, such as the ʻŌlapa *(Cheirodendron trigynum).*

Initial research findings were reported in the *Hawaiian Planters Records* in 1909. The conclusions were published a decade later in the same journal (see Annotated Bibliography, Appendix C, p. 241).

The principal researcher was the botanist Harold Lyon (known also for establishing what is now called the Lyon Arboretum in upper Mānoa valley on Oʻahu). He attacked the Maui forest trouble (as initially dubbed by his colleague Lewton-Brain) as a forest disease. But after a decade of thorough search, no disease organism could be found to be responsible as the killer agent for the ʻŌhiʻa forest dieback on Maui, a finding repeated in the more recent dieback on Hawaiʻi.

Lyon then focused on the soil. It was poorly drained in most places and locally flooded. There was evidence of sulphite bacteria at work, producing occasional oil slick-like puddles. Lyon remarked that he had buried a rusted axe under the shallow root of an ʻŌhiʻa tree (Photo 4.34 and 4.35). After retrieving the axe several months later, he could remove the rust layer by hand to the blank iron. He considered reducing iron as the toxicity agent of the dieback and attributed the toxification to the aging process in Hawaiian soil development.

He referred to the ʻŌhiʻa rainforest as a pioneer forest that cannot adapt to aging volcanic soils. He further reasoned that because of the isolation of the Hawaiian archipelago, the climax species component was missing. These shade-tolerant taller growing climax trees, typically present in the more species rich tropical continental forests, and often producing larger animal dispersed fruits, had simply not arrived.

To save the watersheds in Hawaiʻi, Lyon suggested therefore, to introduce tree species from other tropical areas to re-

Photo 4.34. Dieter Mueller-Dombois holding an ʻŌhiʻa sapling that was pulled out of the organic peat overlying the boggy, alpha gley soil. Mallikarjuna Aradhya is pointing at the lateral roots which developed from the stem of the plant that was growing in accumulated peat, an adaptation by this species to survive on a soggy substrate.

place the pioneer ʻŌhiʻa forest. Subsequently, many tree species were introduced for trial. Lyon favored fig species, because once established, they would not be harvested since they are commercially of low value. However, figs were hard to establish. Instead, he finally used as successful replacers on water soaked soil two tree species, the Australian swamp mahogany (*Eucalyptus robusta*), and the New Caledonian paper bark tree (*Melaleuca quinquenervia*) (Photo 4.36).

Photo 4.35. Bush knife stuck in a drainage channel within the Maui dieback forest terrain. This channel was initiated by feral pigs and further used as a hunter's trail, and may eventually lead to the development of a new stream bed.

Photo 4.36. A 1985 view of the developing ʻŌhi'a shrub bog on windward Haleakalā on the island of Maui with an old ʻŌhi'a tree snag behind Maimuna Morshidi. The Paperbark tree plantation in the background dates from the CCC (Civilian Conservation Corps) days in the 1930s. The swamp adapted Paperbark tree, native to New Caledonia, has since started invading the native habitats adjacent to the plantation.

About half of the Maui dieback area was successfully replanted with these two introduced species during the depression by the CCC (Civilian Conservation Corps). The planted stands of both species were successful also in reducing the sogginess of the soil. In other words, they "mopped up" the forest floor where planted. All seemed successful except for an error in interpretation.

It was not the soil alone that became unsuitable for the ʻŌhiʻa pioneer forest. It was the broader landscape. In this zone of high annual rainfall of 4,000–7,000 mm (158–275 in) on gentle slopes of the Hawaiian shield volcanoes, the former perfectly drained and good forest habitat was changing into boggy habitat. None of the native rainforest tree species is available to grow into tall trees on such boggy habitats. However, ʻŌhiʻa is still able to survive as dwarf or stunted tree or shrub, where it grows anew on its own built-up organic matter and woody peat, including that of fallen tree ferns and associated woody plants.

In other words, bog-formation dieback can be described as a **"forest stand reduction dieback,"** not as total dieoff or ʻŌhiʻa graveyard.

Stream formation

In the bog-formation territory on Mauna Kea we distinguished two types of bogs, organic bogs and clay bogs. Organic bogs were accumulating dead organic matter from a combination of sedge, fern and woody peat on the surface of the water soaked mineral soil. On top of the organic bogs grew the native mat forming Uluhe fern together with stunted or dwarfed ʻŌhiʻa trees. The clay bog vegetation was dominated by the alien Broomsedge (*Andropogon virginicus*).

We noted during rains that surface water was slowly moving over the clay bogs. Following the path of the water revealed that these clay bogs functioned as incipient places of stream development.

Thus, in the bog-formation territory, both on Mauna Kea and Maui, the formerly forested landscape is changing very slowly on a geomorphological time scale into two types of new ecosystems, bogs and streams. These new successional ecosystem types can be considered just as effective in their watershed function as the former ʻŌhiʻa forest that occupied the same location on the slope.

Conclusion

The five types of dieback discovered during our field research gave interesting indications of successional changes. The wetland and the dryland dieback types showed that both could be considered ʻŌhiʻa **replacement dieback** types in the progression phase of primary succession. Their structural change indications contrasted with the 'Ōhi'a **displacement dieback** type in the eutrophic climax phase of the primary succession (or in the chronosequential change of forest habitats).

After the eutrophic peak we encountered the gap-formation and bog-formation types of dieback. In both types, 'Ōhi'a tree cohorts occupied smaller areas. Since these two dieback types were in the regression phase of the chronosequence of rainforest landscapes, their recovery with 'Ōhi'a individuals appeared to be less vigorous. New individuals tended to become smaller trees than those that went into dieback. Based on structural characteristics, the gap- and bog-formation diebacks were seen as **stand reduction diebacks** for 'Ōhi'a trees.

Finally, stream formation beginning on fluvial flats among boggy habitats, show **complete die-off of 'Ōhi'a** trees. But as soon as streams cut their new pathway downslope, 'Ōhi'a seedlings return on the rejuvenated soil of the stream banks, an observation made also in the Maui dieback area by Harold Lyon in the early part of the last century.

Suggested Readings

(See also Appendix C: Research History of the 'Ōhi'a Rainforest Dieback/Decline)

Akashi, Y. & Mueller-Dombois, D. (1995). A landscape perspective of the Hawaiian rainforest dieback. *Journal of Vegetation Science* 6: 449–464.

Auclair, A. N. D. (1993). Extreme climatic fluctuations as a cause of forest dieback in the Pacific rim. *Water Air Soil Poll.* 66: 207–229.

Auclair, A. N. D., Worrest, R. C., Lachance, D. & Martin, H. C. (1992). Climatic Perturbation as a Mechanism of Forest Dieback. In *Forest Decline Concepts*. ed. by Paul D. Manion and Dennis Lachance. St. Paul, MN: APS Press. pp. 38–58.

Balakrishnan, N. & Mueller-Dombois, D. (1983). Nutrient studies in relation to habitat types and canopy dieback in the montane rain forest ecosystem, island of Hawai'i. *Pacific Science* 37(4): 339–359.

Burgan, R. E. & Nelson, R. E. (1972). *Decline of 'Ōhi'a Lehua Forests in Hawaii*. Berkeley, CA: US Department of Agriculture, Forest Service, Pacific SW Forest 6 Range Expt. Stn., General Technical Report PSW-3. 4 p.

Burton, P. J. (1982). The effect of temperature and light on *Metrosideros polymorpha* seed germination. *Pacific Science* 36(2): 229–240.

Burton, P. J. & D. Mueller-Dombois (1984). Response of *Metrosideros polymorpha* seedlings to experimental canopy opening. *Ecology* 65(3): 779–791.

Ciesla, W. M. & Donaubauer, E. (1994). *Decline and Dieback of Trees and Forests: A Global View*. Rome: Food and Agriculture Organization of the United Nations, FAO Forestry Paper 120. 90 p.

Clarke, F. L. (1875). Decadence of Hawaiian forest. In *Thrum's Hawaiian Annual, All About Hawaii.* Vol. 1. Honolulu: Star-Bulletin Printing Co. pp. 19–20.

Cooray, R. G. & Mueller-Dombois, D. (1981). Feral pig activity. In *Island Ecosystems: Biological Organization in Selected Hawaiian Communities*, ed. by D. Mueller-Dombois, R. W. Bridges, & H. L. Carson. Stroudsburg, PA: Hutchinson Ross Publishing Company. pp. 309–317.

Cox, G. W. (1999). *Alien Species in North America and Impacts on Natural Ecosystems.* Washington, D C: Covelo, Island Press. 387 p.

Drake D. R. & Mueller-Dombois, D. (1993). Population development of rain forest trees on a chronosequence of Hawaiian lava flows. *Ecology* 74: 1012–1019

Drake, D. R. & Pratt, L. W. (2001). Seedling mortality in Hawaiian rain forest: The role of small scale physical disturbance. *Biotropica* 33(22): 319–323.

Evenson, W. E. (1983). Climate analysis in ʻŌhiʻa dieback area on the island of Hawaiʻi. *Pacific Science* 37(4): 375–384.

Gerrish, G. C., Mueller-Dombois, D. & Bridges, K. W. (1988). Nutrient limitation and *Metrosideros* dieback in Hawaiʻi. *Ecology* 69(3): 723–727.

Hodges, C. S., Adee, K. T., Stein, J. D., Wood, H. B. & Doty, R. D. (1986). *Decline of ʻŌhiʻa (*Metrosideros polymorpha*) in Hawaii: A Review*. Berkeley: US Department of Agriculture, Forest Service, Pacific Southwest Forest and Range Experiment Station. General Technical Report PSW-86. 22 p.

Holt, R. A. (1983). *The Maui Forest Trouble: A Literature Review and Proposal for Research.* Honolulu: University of Hawaiʻi, Hawaiʻi Botanical Science Paper No. 42. 67 p. URL: www.botany.hawaii.edu/pabitra.

Huettl, R. F. & Mueller-Dombois, D. (eds.) (1993). *Forest Decline in the Atlantic and Pacific Regions.* Third Intern. Dieback Symposium. Heidelberg: Springer-Verlag. 366 p.

Kliejunas, J. T. & Ko, W. H. (1973). Root rot of ʻŌhiʻa (*Metrosideros collina* ssp. *polymorpha*) caused by *Phytophthora cinnamomi*. *Plant Disease Reporter* 57: 383–384.

Kliejunas, J. T. & Ko, W. H. (1974). Deficiency of inorganic nutrients as a contributing factor to ʻŌhiʻa decline. *Phytopathology* 64: 891–896.

Lewton-Brain, L. (1909). The Maui forest trouble. *Hawaiian Planter's Record* 1: 92–95.

Lyon, H. L. (1909). The forest disease on Maui. *Hawaiian Planter's Record* 1: 151–159.

Lyon, H. L. (1919). Some observations on the forest problems of Hawaiʻi. *Hawaiian Planter's Record* 21: 289–300.

Manion, P. D. (1981). *Tree Disease Concepts*. Englewood Cliffs, NJ: Prentice Hall. 399 p. 1991 Second ed. 402 p.

Manion, P. D. & Lachance, D. (eds.) (1992). *Forest Decline Concepts*. St. Paul, MN: APS (American Phytopathological Society). 249 p.

Mueller-Dombois, D. (1992). A natural dieback theory, cohort senescence as an alternative to the decline disease theory. In *Forest Decline Concepts*, ed by P. D. Manion and D. Lachance. St. Paul, MN: APS Press. pp. 26–37.

Mueller-Dombois, D. (1993). Biotic impoverishment and climate change: Global causes of forest decline? In *Forest Decline in the Atlantic and Pacific Regions*, ed. by R. F. Huettl & D. Mueller-Dombois. Springer-Verlag Berlin, Heidelberg: Springer-Verlag. pp. 338–348.

Mueller-Dombois, D., Bridges, K. W. & Carson, H. L. (eds.) (1981). *Island Ecosystems: Biological Organization in Selected Hawaiian Communities*. Stroudsburg, PA & Woods Hole, MA: Hutchinson Ross Publishing Company, US/IBP Synthesis Series 15. 583 p.

Mueller-Dombois, D., Jacobi, J. D., Cooray, R. G. & Balakrishnan, N. (1980). *ʻŌhiʻa Rain Forest Study: Ecological Investigations of the ʻŌhiʻa Dieback Problem in Hawaiʻi.* Honolulu: University of Hawaiʻi, College of Tropical Agriculture and Human Resources, Hawaiʻi Agricultural Experiment Station, Miscellaneous Publication 183. 64 p.

Papp, R. P., Kliejunas, J. T., Smith, R. J. & Scharpf, R. F. (1979). Association of *Plagithmysus bilineatus* and *Phytophthora cinnamomi* with the decline of ʻŌhiʻa lehua forests on the island of Hawaii. *Forest Science* 25: 187–196.

Petteys, E. Q. P., Burgan, R. E. & Nelson, R. E. (1975). *ʻŌhiʻa forest decline: Its spread and severity in Hawaii.* Albany, CA: USDA Forest Service, PSW-105. 11 p.

Podger, F. D. (1981). Definition and diagnosis of diebacks. In *Eucalypt Dieback in Forests and Woodlands*, ed. by Old, K. M., Kile, G. A. & C. P. Ohmart. Melbourne: CSIRO. pp. 1–8.

Wargo, P. M. & Auclair, A. N. D. (2000). Forest declines in response to environmental change. In *Responses of Northern US Forests to Environmental Change*, ed. by R. A. Mickler, R. A. Birdsey & J. Hom. Berlin, Heidelberg: Springer-Verlag, Ecological Studies 139. pp. 117–145.

Chapter 5

Dieback as a Natural Process in Succession

In a final, very thorough review, published in 1986 by a team of five US Forest Service Researchers, the collapse in the ʻŌhiʻa rainforest was interpreted as a "**typical decline disease**" (C. S. Hodges, K. T. Adee, J. D. Stein, H. B. Wood & R. S. Doty 1986).

A disease is an illness or abnormality. Since the ʻŌhiʻa collapse or dieback was definitely proven not to be caused by a newly introduced biotic agent specifically, it cannot be correctly interpreted as a decline disease. Neither can it be interpreted simply as a physiological disease. While we found nutritional imbalances associated with the change in soil and landscape aging from the younger Mauna Loa to the older Mauna Kea volcanoes, these are normal and completely natural changes in forest habitats within the Hawaiian Islands that occur over a time frame of approximately one million years.

The soil nutrient composition changes from eutrophic (nutrient-rich), as found in areas with ʻŌhiʻa displacement dieback, to oligotrophic (nutrient-poor), as in the gap and bog-formation dieback types are completely natural. These are normal changes associated with the aging of landscapes on these Hawaiian mountains; therefore, the interpretation of ʻŌhiʻa dieback has to be a natural or normal phenomenon.

A long-term primary succession

The conceptual model shown in Fig. 5.1 summarizes and reflects some of the research results about the rainforest dieback discovered in a successional context. There are the two common volcanic substrates differentiated in this figure: pāhoehoe lava forming the lower curve and volcanic ash forming the upper curve. Volcanic ash includes all pyroclastic (fire-broken) substrates including 'a'ā flows and cinder deposits. Pyroclastic substrates offer nutrient advantages for 'Ōhi'a forest development. Volcanic ash and cinder change more readily into soil than solid rock surfaces, and 'a'ā lava allows for easier root penetration than does the pavement-like pāhoehoe. The pyroclastic substrates are thus superior in terms of nutrient availability for the plants growing on them. Such substrates result in more plant biomass in the rainforests as compared to the pāhoehoe substrates (as indicated by the different heights of the curves and on the upright axes or Y-axes on Fig. 5.1). The process of substrate development toward enhanced nutrient availability takes about 1,000 to 10,000 years as indicated by the peak (or climax) on the diagram. This is the **progression phase** in primary succession, which includes two of the dieback types, the 'Ōhi'a displacement and dryland dieback. Dryland dieback also occurred on substrates of volcanic ash and 'a'ā lava.

A regression phase follows the nutrient-rich peak as minerals essential for plant growth become less available. This phase includes the three other recognized dieback types, the wetland dieback on pāhoehoe, the gap-formation dieback on soils from volcanic ash, and finally the bog-formation dieback on soils from pyroclastic parent material but also from deeply weathered pāhoehoe.

Note that the horizontal (or X-axis) on Fig. 5.1 has a logarithmic scale. This implies that the regression phase along the chronological sequence of primary succession takes very much longer (from 10,000 to 1,000,000 years) than the progression phase. The age limits and ranges on Fig. 5.1 are only approximations along the timeline of ecosystem and landscape aging in Hawai'i. However, not all forest landscapes in Hawai'i end in bogs; most areas develop into stream/ridge systems with new soil exposed as a result of erosion.

Secondary successions in form of dieback cycles

Along the primary succession curves on Fig. 5.1 you can see abrupt decreases in forest biomass with gradual, but relatively short time, recovery phases. These checkmark like symbols indicate periodically repeating dieback cycles followed by secondary successions with forest regeneration. The primary succession time line shows that following canopy dieback and loss of live tree biomass, the forest again increases in biomass as it recovers from this event. In the aggrading phase of succession

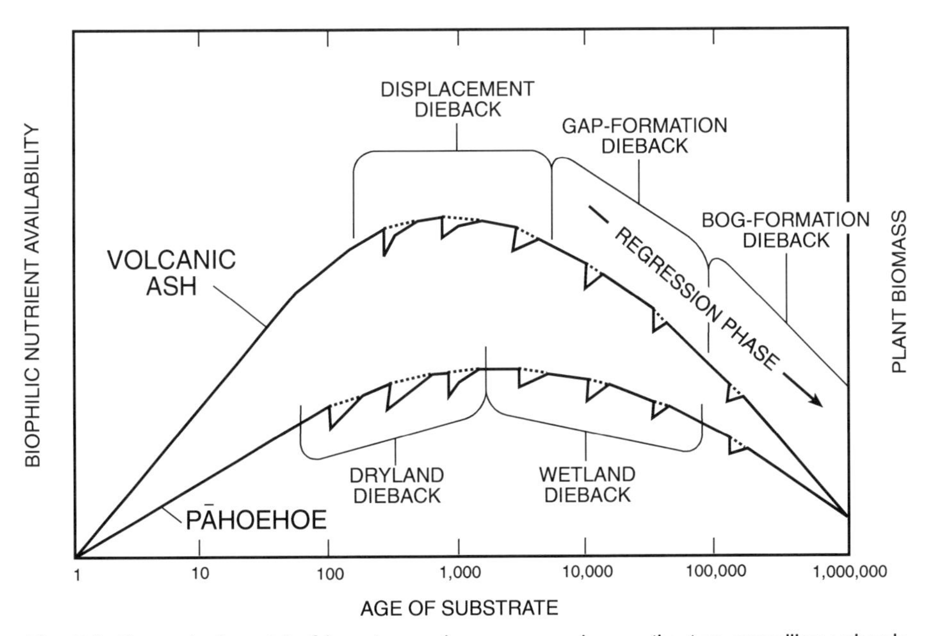

Fig. 5.1. Conceptual model of long-term primary succession on the two prevailing volcanic substrates in Hawai'i with the five dieback types superimposed (reproduced with permission from *Annual Review of Ecology & Systematics* 1986: 234).

the biomass returns to, at least, its previous level. However, in the regression phase (right side of the graph), each successive forest generation following canopy dieback recovers to less plant biomass that what it had prior to dieback. In other words, the overall forest is declining (Further detailed in Chapter 6).

Soil formation

Soil research revealed stress factors, such as nitrogen deficiency on young volcanic soils. Excess soil moisture was seen as causing wetland dieback by flooding of the root system resulting in lack of oxygen for root respiration. On old soils, deficiencies of elements, such as magnesium and calcium, were apparent. When combined with poor drainage as in the bog-formation dieback, soil toxicity from reduced iron, as well as excessive manganese and aluminum became important stress factors. Phosphorus is the most limiting nutrient in old volcanic soils.

The progression of nutrient status over the course of volcanic soil development in Hawai'i was summarized by Fox et

THE LONG-TERM CHRONOSEQUENCE

Ecological research across the high island chain, from the young island Hawai'i to the 5 million year old Kaua'i, has shown that not all rainforest development ends in bogs.

Research sites in 'Ōhi'a rainforests, all located at 1,200 m (4,000 ft) elevation and in the same rainfall regime of 2,500 mm/year (~100 in/year) terminate on Kaua'i in low stature forests with only few Hāpu'u tree ferns in the undergrowth, but not in boggy habitats. In contrast, research sites at the same elevation but in much higher rainfall regime of 4,000 mm/year (~160 in/year) do terminate in boggy habitats with permanently soggy soils. They are characterized by having a proportionately higher dead than living biomass.

See Bibliography for more details by Kitayama & Mueller-Dombois (1995) and by Vitousek (2004).

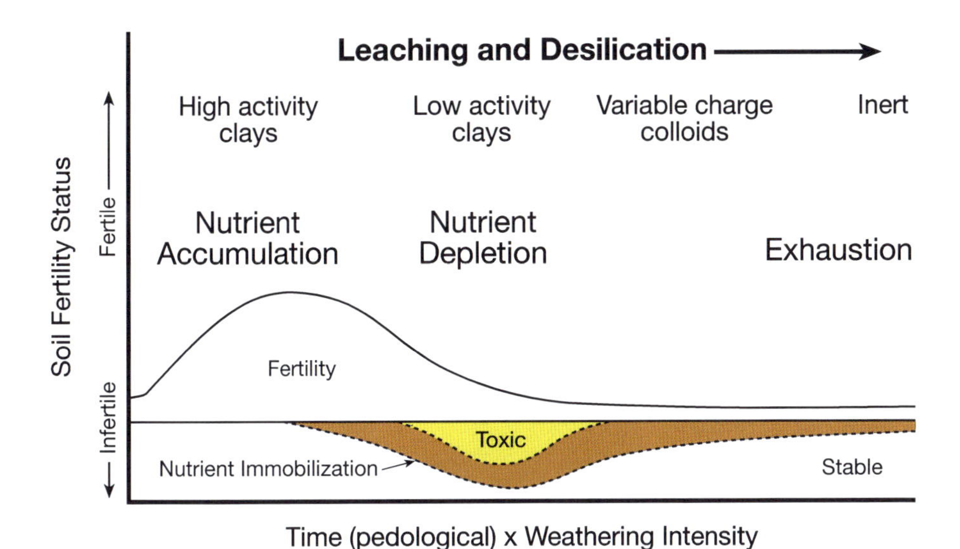

Fig. 5.2. Tropical soil formation and degradation, an independently designed conceptual model that matches the long-term primary sucessional model shown in Fig. 5.1 above. From Fox et al. 1991 with permission from *Allertonia*, a journal of the National Tropical Botanical Garden on Kaua'i Island.

al. (1991) (Fig. 5.2). The nutrient accumulation and nutrient depletion phases up to the toxic soil status coincide with the one million year time series of progressive and regressive forest succession shown on Fig. 5.1.

Historic evidence of dieback and recovery cycles

Fig. 5.3 shows a historic perspective of collapse followed by recovery of ʻŌhiʻa rainforest around a bog on the Island of Molokaʻi over the past 10,000 years. This record of fluctuating pollen abundance is based on percent with other pollen including tree fern spores. The latter often seemed to overlap in abundance with the ʻŌhiʻa pollen thus indicating the ʻŌhiʻa displacement type of dieback. The oldest carbon date of the core points to the start of the bog 10,000 years ago, which according to Fig. 5.1 was in the aggrading phase of primary succession. Whatever meaning one may attribute to the pollen core, it reflects the long-term resilience of ʻŌhiʻa as an **oscillating persister** in

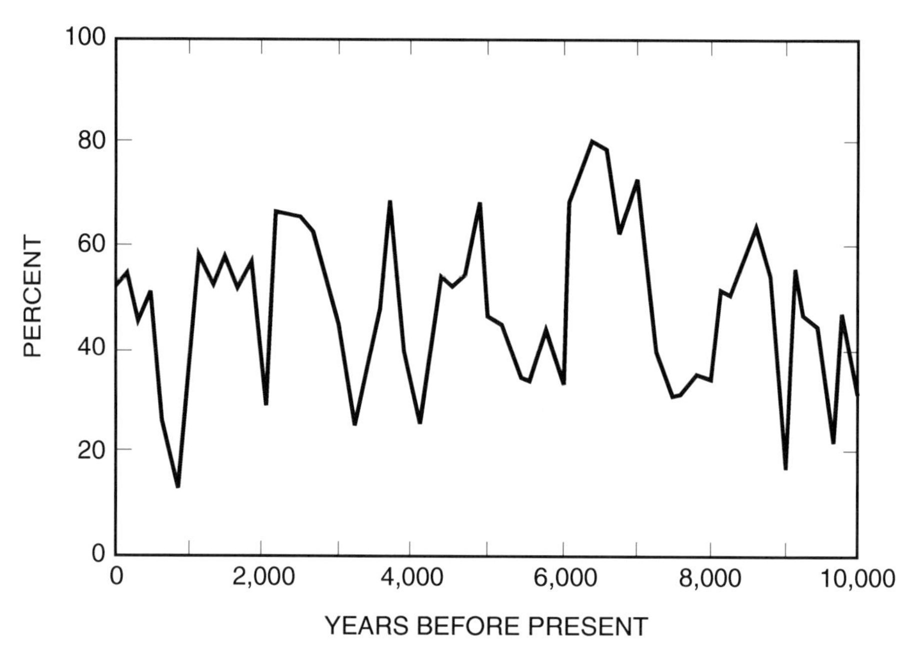

Fig. 5.3. The fluctuating abundance of 'Ōhi'a tree pollen in a 325 cm (10.7 ft) deep soil core in a Hawaiian rainforest bog on Moloka'i (after Selling 1948). The time axis is based on carbon dating the core base. (Reproduced with permission from *Annual Review of Ecology & Systematics* 1986: 233).

the Hawaiian rainforest. The two to three pollen depressions per 1,000 years, indicated on the diagram, suggest that the duration of 'Ōhi'a tree stand generations may be associated with canopy dieback or possibly other more drastic disturbances, such as hurricanes.

Geomorphological aging of a Hawaiian shield volcano

Figure 5.4 shows an even longer time scale of landscape aging. It takes about 2 million years on the rainy windward sides of the Hawaiian Islands for the original shield to break down.

Here, obviously the original windward 'Ōhi'a forest habitats, including whatever bogs may have been formed marginally after them, have long been eroded away many times. Erosion in form of submarine landslides, sheet erosion, and relentless

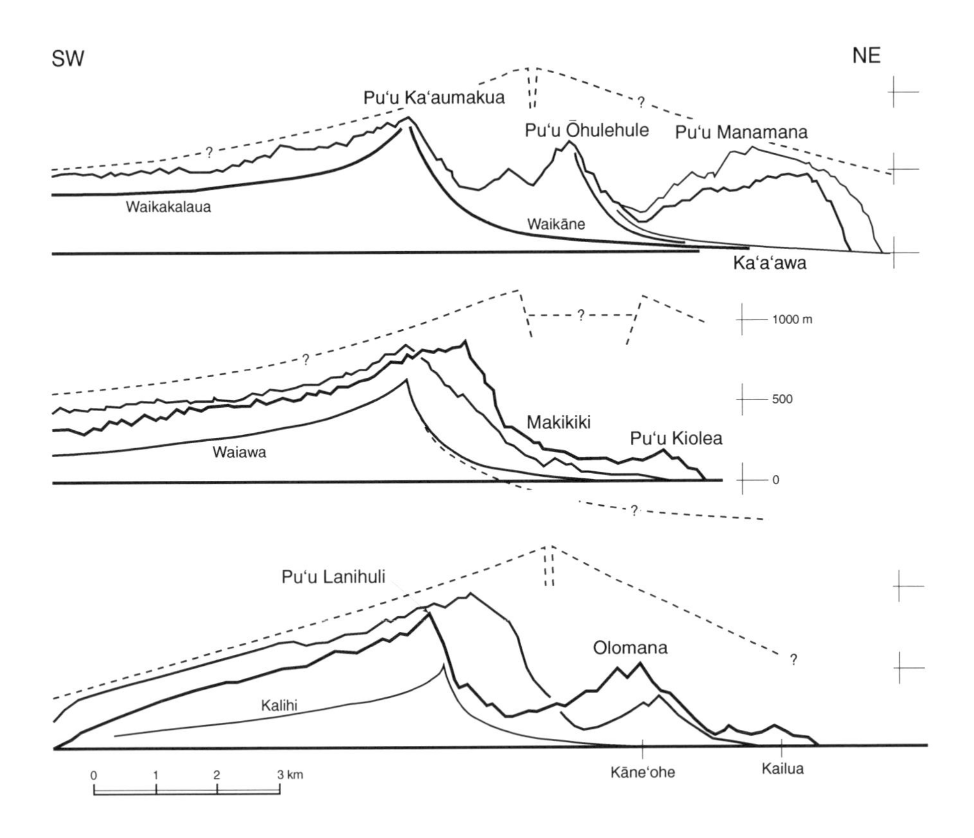

Fig. 5.4. Three profiles through east O'ahu island showing the geomorphic change over 2 million years from erosion of the Ko'olau shield volcano. (Diagram from Wirthmann & Hueser 1987 with permission from *Geographische Rundschau*).

(but normal) soil erosion has resulted in a new landscape and ecosystem mosaic. One can also view this as a rejuvenation of landscapes that follows over a very long-term period of time.

Three potential explanations for 'Ōhi'a dieback

The first cause proposed to explain the 'Ōhi'a dieback/decline syndrome was thought to involve a killer disease. The US Forest Service Team had thoroughly clarified by 1980 that a biotic disease could not be made responsible. Thus there was no newly introduced biological killer agent involved in directly causing the 'Ōhi'a dieback.

WHAT IS A DECLINE DISEASE?

In a forest pathology textbook, Paul D. Manion, an internationally well-known and respected forest pathologist, recognizes three types of diseases: (1) Biotic diseases, a clear case of tree death caused by a disease or pest organism, (2) Physiological diseases referring to tree death caused by abiotic/environmental stresses, and (3) Decline diseases involving a syndrome of stress factors. These may include competition, suppression, defoliation, fungi and bark beetles, all working together in sequential stages. This chronic stress sequence is conceptualized in form of a death spiral. It thus ends up in finally killing the affected tree. In other words, there is a chronological chain reaction of causes, which are very difficult to untangle.

The second factor complex thought causing the dieback was a combination of **abiotic stress factors**. Research focused on the hypothesis that "The dieback is initiated by a climatic instability which becomes effective through the soil moisture regime under certain conditions of forest maturity" (Mueller-Dombois 1980).

Several analyses of climate change and/or disturbances were done, including stream flow and water quality analyses. No excessive stream flow or mudding of the water coming down from the Mauna Kea dieback area was detected. Stream flow was also not reduced. Nothing abnormal could be detected. The same was noted in records of sugar cane plantations during the investigation of the Maui rainforest dieback early in the last century (Holt 1983).

Since no loss of watershed value was found in association with forest dieback, it became particularly clear that the foliage loss and crown death was restricted to canopy trees, while the undergrowth remained healthy. An exception was seen in the bog-formation dieback, where the undergrowth was also affected, for example the Hāpu'u tree ferns in these areas were dying as well. This process we called "stand reduction dieback." But bog-formation on the lower slopes of the two shield volcanoes (Mauna Kea on Hawai'i Island and Haleakalā on Maui) also led to regulating the water flow just as forested habitats were regulating the water flow in their place before. Moreover, bog-formation is also associated with stream formation, which is a slow and very gradually emerging process of surface drainage. The emerging streams provide initially for a very gradual water flow down slope, while the bogs act like kidneys, storing and releasing water slowly to the emerging streams.

Since neither biotic disease factors nor abiotic stress factors could explain the forest collapse, it became apparent that a third factor complex must be at work in the dieback/decline syndrome.

The 1980 working hypothesis (stated above) hinted toward a third factor complex by including the phrase "under certain conditions of stand maturity."

The third explanation: Succession of life stages

This third explanation of 'Ōhi'a dieback involves a complex set of factors related to the history and demography of 'Ōhi'a forest stands. From the origin of such stands on new volcanic surfaces, we learned that 'Ōhi'a seedlings tend to become established frequently in cohorts over large areas.

Here they form stands belonging to the same generation. This invasion process is often uneven on the same lava flow or cinder deposit. Thus, there can be groups of earlier and later arrivals. Also within groups there are early and late arrivals. Some early colonizers produce flowers and seeds, which add new younger plants to the site. Thus, the new generation is not a single "birth cohort" such as found in annual plants.

Annual plants grow up from seed, often in form of birth cohorts. When mature, they flower and produce seeds. After seed shedding, they die. Their life cycle is typically completed in one year. Not so in perennial plants, such as the 'Ōhi'a trees. They also have a life cycle, which is genetically programmed. No one, as we know, has yet attempted to determine that genetic program of life expectancy. But we know that trees like all other sexually reproduced organisms have a life cycle. Trees, such as 'Ōhi'a, demonstrate their life cycle in terms of cohorts on the new volcanic surfaces in Hawai'i, as already discussed and shown as photos in Chapter 2. First there is the seedling stage, followed by the sapling stage, the juvenile stage, the mature stage, and old-growth stage. All these stages can be seen in form of generation or cohort stands. In the old-growth stage, the synchrony of cohort individuals may become less apparent, because individual differences in life span can show up in less stressful, nutritionally favorable, habitats. However, under normal conditions there is a senescent life stage prior to death.

In *Webster's Dictionary* "senescence" is defined for plants as "The growth phase from full maturity to death that is characterized by an accumulation of metabolic products, increase in respiratory rate and loss in dry weight especially in leaves and fruit." We may add that trees senesce naturally when their photosynthetic apparatus in the form of leaf area declines permanently relative to their respiring non-green living apparatus. The latter is primarily the tree's root system. This process can be measured by the tree's carbon balance. When the ratio of outgoing carbon dioxide through respiration exceeds the incoming carbon dioxide through photosynthesis permanently and relentlessly, the trees are in their senescing life phase. This can be a long enduring life stage.

The theory of cohort senescence

This theory evolved as an alternative to the decline disease theory during our research in the early 1980s. It implies that there is a **third factor complex** to explain 'Ōhi'a tree canopy collapse in the Hawaiian rainforest. The cohort senescence theory includes several elements.

As mentioned before, the cohort senescence theory focuses on the life history of a forest stand. In the beginning, there is a disturbance that opens up a site for a new cohort to become established. To survive as a cohort stand, the individual members must adjust to the site. Some sites are favorable, others are less favorable. In all cases there are stress factors that appear at

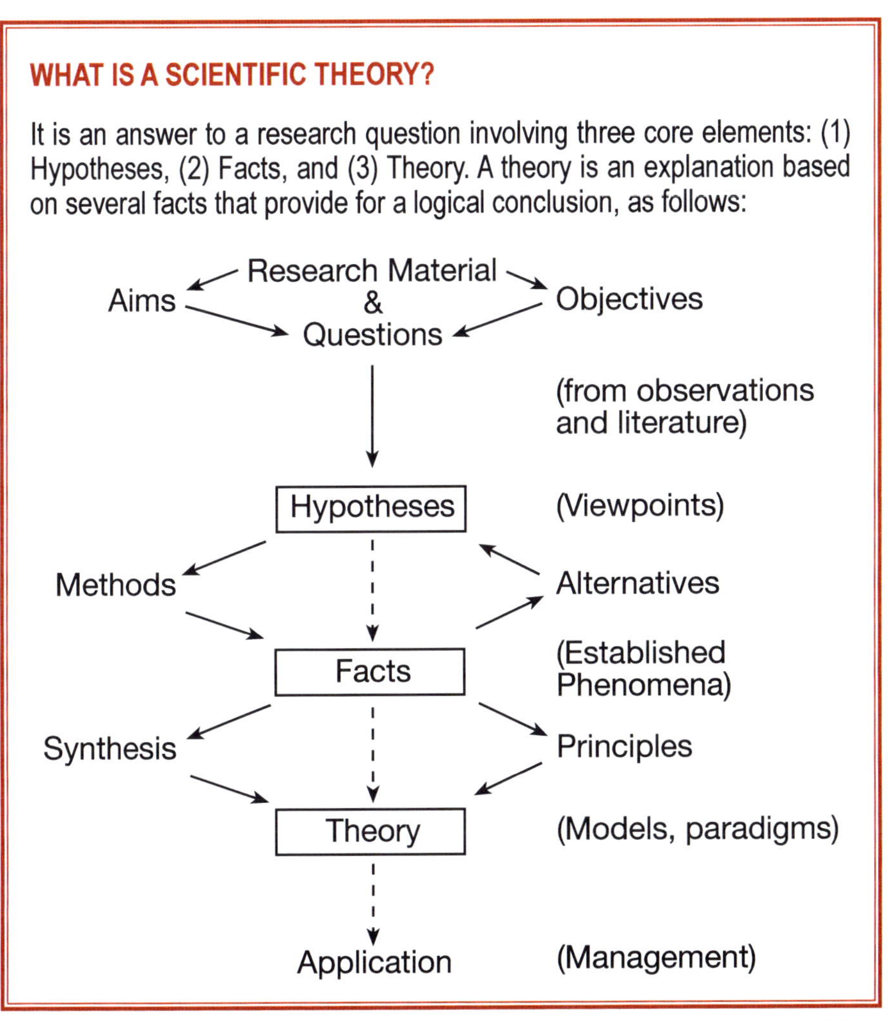

Fig. 5.5. A general research procedure for vegetation ecology. (From *Can. J. Bot.* 66: 2622)

various stages in the cohort's life cycle. Some sites allow cohort individuals to reach their optimum, others present constraints. On sites with constraints a cohort may enter a stage of premature senescence.

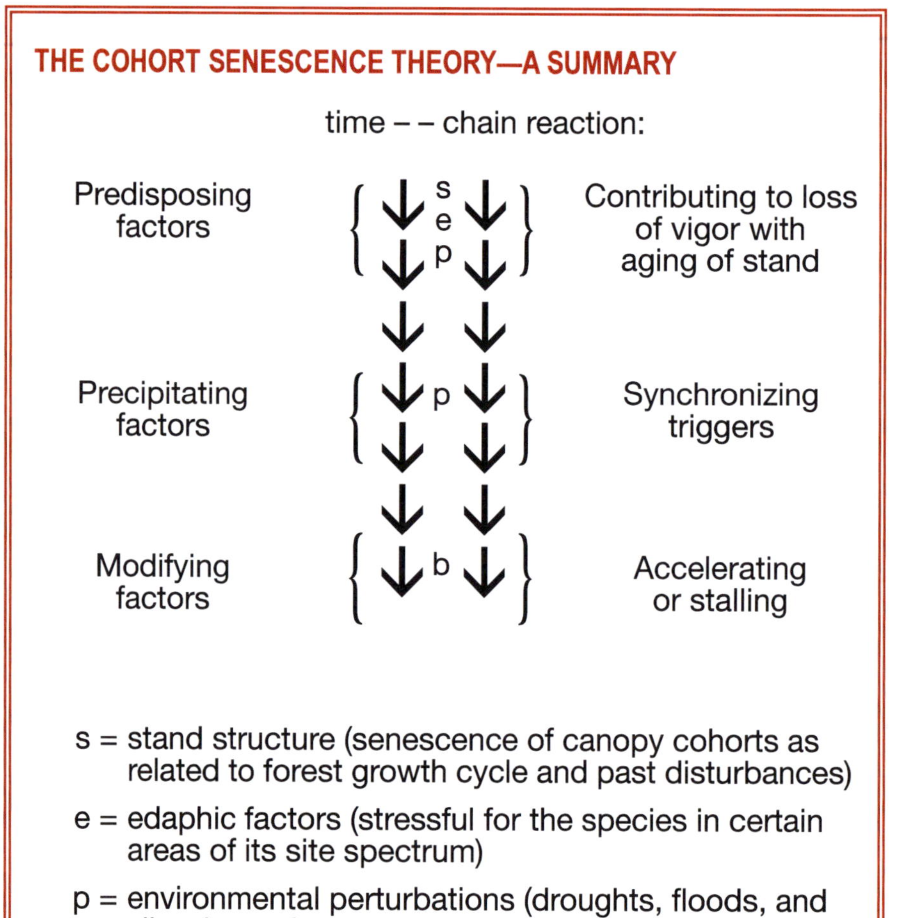

Fig. 5.6. A chain reaction of factors involved in natural canopy dieback.

Facts supporting the cohort senescence theory for ʻŌhiʻa dieback:

(1) Dieback was restricted to the ʻŌhiʻa canopy, while seedlings were unaffected.

(2) Dieback stands were found to be separated from non-dieback stands along substrate boundaries.

(3) Dieback stands were also separated from non-dieback stands on the same substrates.

(4) Dieback occurred on well drained and poorly drained substrates (wetland vs. dryland dieback).

(5) Based on diameter measurements, the ʻŌhiʻa canopy trees tended to form a normal (bell-shaped) size-frequency curve. Such curves are typical for even-aged mature plantation stands. Tropical forests with many canopy species typically have inverse J-shaped diameter curves.

(6) The typical structure of ʻŌhiʻa forest stands showed two cohorts, a canopy cohort and a seedling cohort. As a rule a sapling cohort was absent since ʻŌhiʻa is for the most part shade-intolerant. In other words, mature ʻŌhiʻa stands with closed canopies typically have a **sapling gap**.

Cohort senescence can be visualized as three phases of factors forming a chain reaction:

Predisposing factors, which contribute to loss of vigor with aging of the cohort stand.

Precipitating factors, which act as synchronizing triggers.

Modifying factors, which can accelerate or stall the senescing process.

Predisposing factors include:

s = Stand structure. This factor appears to be of prime importance. If mature stands form a canopy cohort, they become naturally predisposed with advancing age (normal senescence)

e = Extreme edaphic or soil-substrate impediments can predispose stands to dieback, especially where their habitats have low buffering capacities (premature senescence)

p = Periodically recurring climatic disturbances, including heavy rain storms or transitory droughts can weaken a forest stand. Such events may result in a synchronous premature senescence of the individuals in the stand.

Precipitating factors:

These are the same as "p" above. Periodically recurring climatic perturbations or strong weather disturbances can act as **trigger** factors when stands are predisposed to dieback.

Modifying factors:

Finally, naturally occurring biotic agents "b" can give the dying stand the *coup de grace.* This can result in accelerating the dieback. Absence of biotic agents can slow the dieback process. Both the fast and slow forms have been observed in the ʻŌhiʻa dieback/decline syndrome.

Accelerated dieback has positive effects in the ʻŌhiʻa rainforests. Canopy openings release existing shade-born seedlings and favor the emergence of new light-born ones. The light-born seedlings grow faster and survive better following widespread canopy loss.

In the cool wet mountains of Hawaiʻi, dead trees can remain standing for decades due to the slow activity of decomposers. Termites are not native to Hawaiʻi and the introduced species of termites have so far remained below 1,200 m (4,000 ft) in Hawaiʻi.

The climatic trigger factor: A hypothesis

Figure 5.7 shows a graph indicating extreme rainfall events for Hilo, Hawai'i from 1900–1984. Asterisks on the left of the midline indicate severe droughts, less than 10 mm (0.4 in) per month or two or more sequential months with less than 50 mm (2 in) rain. Asterisks on the right indicate severe floods, with over 1250 mm (50 in) rain/month or two more sequential months with more than 750 mm (30 in). The shaded zigzag line in the center of Fig. 5.7 represents annual rainfall divided by 12 for each individual year from 1900 to 1984.

Climatic disturbances are indicated by those events that extend beyond the dashed vertical index lines. Crown wilting has been noted on rainforest trees when rainfall was less than 50 mm/month (*drought index*), and prolonged impounding of surface water was seen on poorly drained soils when rainfall exceeded 750 mm/month (*flood index*). There were two extreme wet months in 1901 and 1902 followed by a concentration of extreme drought events early in the 20th century (March 1904, January 1905, February and March 1906). These may have precipitated or acted as trigger of the Maui bog-formation dieback, which was discovered in 1905. Heavy storms may have been associated with these extreme wet weather events. That forest was certainly predisposed to die. Similarly, we can recognize cumulative flood years from 1921 to 1923, which may have further aggravated the Maui dieback.

However, this analysis primarily focuses on the rainforest climate near Hilo, which is centrally located below the forest dieback area on the Island of Hawai'i. Thus, the same years (1921, 1922, and 1923) may have triggered the Mauna Kea bog-formation dieback. The other dieback types may have been initiated in January 1953, when there was an extreme drought followed by an extreme flood event in December 1954. This hypothesis may be testable with diurnal weather analyses and field

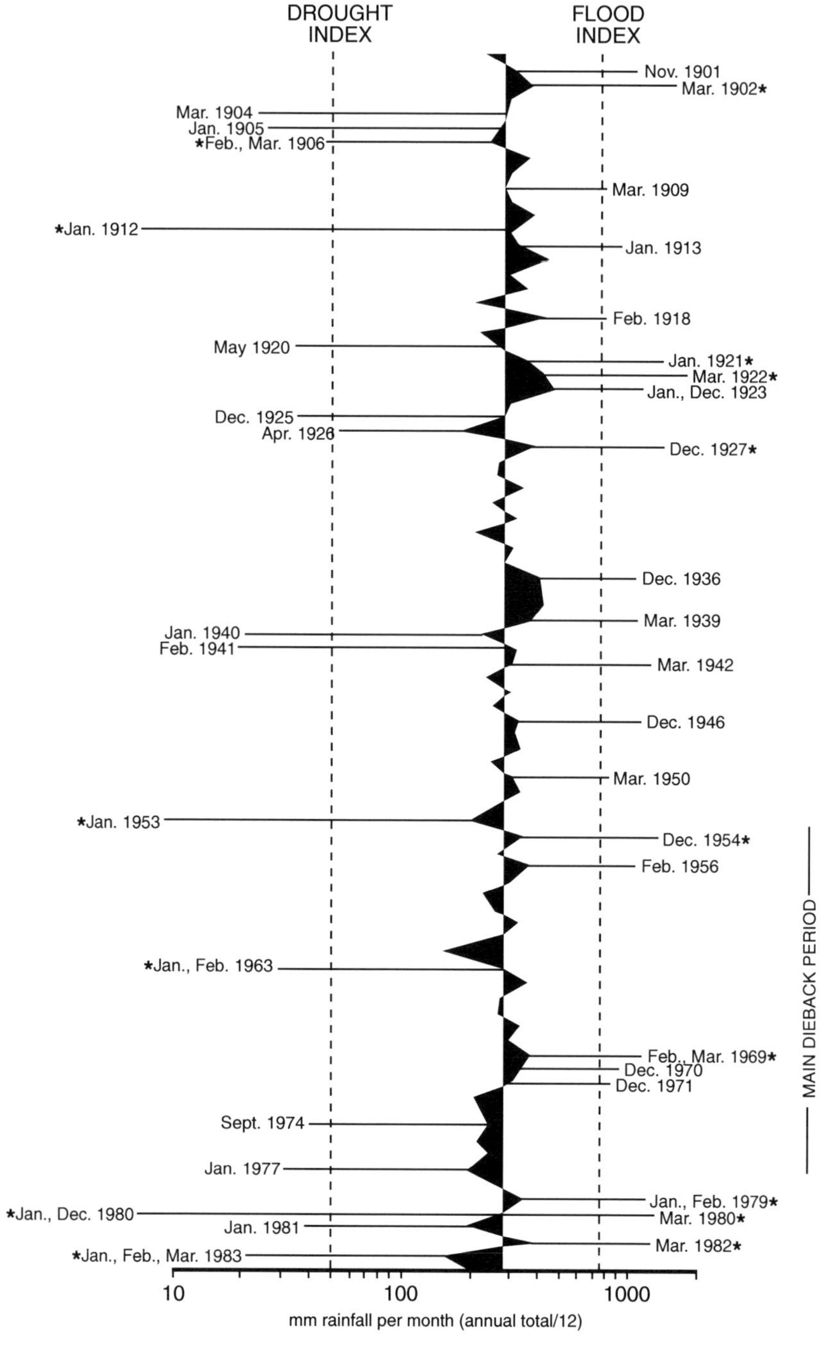

Fig. 5.7. Extreme rainfall months at Hilo airport from 1900 to 1984. Months with < 50 mm (2 in) rainfall are shown left of drought index and those with > 750 mm (30 in) at right of flood index. Months marked with an asterisk (*) indicate intense drought or flood events.

observations by monitoring tree stands that are in a low vigor state and thus appear to be predisposed to enter into dieback phase. Physiological measurements can be used to ascertain a tree's vitality status.

During the three main dieback decades across Mauna Kea and Mauna Loa (from 1954 to 1984, when 50% of a 100,000 hectare rainforest territory experienced dieback), we see an accumulation of extreme climatic disturbances. We cannot isolate, as yet, a single climatic disturbance event or a single trigger factor from this analysis. However, further monitoring of drought

MORE DETAIL RELATING TO FIGURE 5.7

Hilo airport is the only station with continuous long-term records. This station is located near sea level and centrally below the former dieback area.

(1) The straight vertical centerline represents the long-term annual rainfall at Hilo airport divided by 12, i.e., 295 mm (12 in). This simple reduction is reasonable because mean monthly variations in rainfall at Hilo are not strongly seasonal.

(2) The shaded zigzag line in the center is the year-to-year rainfall from 1900 to 1984 divided by 12. The purpose is to compare the mean annual rainfall to the mean monthly extremes, which are represented by the extended horizontal lines.

(3) The dashed vertical lines denote arbitrary threshold values. A FLOOD INDEX is shown on the right (wet) side at 750 mm/month (30 in). This implies temporary impounding of water on poorly drained surfaces whenever the monthly rainfall exceeds 750 mm. A DROUGHT INDEX is shown on the left (dry) side at 50 mm/month (2 in). Canopy foliage in rainforests has been noted to wilt whenever the monthly rainfall is less than 50 mm.

(4) The horizontal lines crossing the threshold values show the oscillation of extreme rainfall months. Note, that the horizontal axis at the bottom of Fig. 5.7 has a logarithmic scale.

events and similar short-term events of storms that resulted in major losses of canopy foliage would further clarify the nature of 'Ōhi'a canopy trigger factors.

Results show that the number of months with droughts and floods did not increase from 1900–1984 (rather the opposite 9 versus 7 and 14 versus 8).

However, the intensity of drought events as well as flood events increased during the main dieback period as shown by the percent increases from the first 5 decades into the following 3 decades. This repeated intensification may explain the domino style collapse of half of the 100,000 ha native rainforest terrain over the main dieback period.

SUMMARY OF DIEBACK TRIGGER ANALYSIS FROM 1900–1984
as read from the Diagram of Hilo Rainfall Data Fig. 5.7

(1) Intense Droughts / Drought frequency

1900-1950 = 50 years 2/9 = 22% intense droughts
1951-1984 = 34 years 4/7 = 57% intense droughts

(2) Intense Floods / Flood frequency

1900-1950 = 50 years 4/14 = 29% intense floods
1951-1984 = 34 years 5/8 = 63% intense floods

Conclusions

The analysis of extreme weather events in conjunction with a search for dieback triggers that possibly initiated the Hawaiian forest decline in the 1900s revealed that the frequency of extreme events had increased on windward Hawai'i from 1900 to 1984. Extreme drought periods increased from two in the first 50 years to four in the following 34 years. Likewise, the record showed four extreme storm-flood periods during the first

50 years and five during the next 34 years. The data were too scanty for attaching any statistical significance to them. However, a trend of increasing frequency and intensity of climatic pertubations was indicated (see Fig. 5.7). This increase in the oscillation of weather extremes may be associated with global warming.

We are certain that the major underlying factor complex for the dieback/decline syndrome is to be found in the predisposition of the canopy trees to die across the landscape over a relatively short period of time. Since many of the trees in these areas became established as cohorts, there is an internal synchronizing factor associated with the dieback stands. When forest stands are predisposed to die, there is only a trigger factor or factor complex required to initiate the collapse. This initiating factor, we suggest, is in the form of climatic extreme events. Since the above analysis has not given a clear answer, we can only refer to the climatic trigger as a hypothesis at this time.

Suggested Readings

Burton, P. J. (1982). The effect of temperature and light on *Metrosideros polymorpha* seed germination. *Pacific Science* 36(2): 229–240.

Burton, P. J. & D. Mueller-Dombois (1984). Response of *Metrosideros polymorpha* seedlings to experimental canopy opening. *Ecology* 65(3): 779–791.

Fox, R. L., de la Pena, R. S., Gavenda, R. T., Habte, M., Hue, N. V., Ikawa, H., Jones, R.C., Plucknet, D. L., Silva, J. A., & Soltanpour, P. (1991). Amelioration, revegetation, and subsequent soil formation in denuded bauxitic materials. *Allertonia* 6(2): 128–184.

Holt, R. A. (1983). *The Maui Forest Trouble: A Literature Review and Proposal for Research.* Honolulu: University of Hawai'i, Hawai'i Botanical Science Paper No. 42. 67 p. URL: www.botany.hawaii.edu/pabitra.

Kitayama, K. & Mueller-Dombois, D. (1995). Vegetation changes along gradients of long-term soil development in the Hawaiian montane rainforest zone. *Vegetatio* 120: 1–20.

Mueller-Dombois, D. (1986). Perspectives for an etiology of stand-level dieback. *Annual Review of Ecology and Systematics* 17: 221–243.

Mueller-Dombois, D. (1988). Community organization and ecosystem theory. *Canadian Journal of Botany* 66: 2620–2625.

Selling, O. H. (1948). Studies in Hawaiian pollen statistics. In *Part III: On the Late Quaternary History of the Hawaiian Vegetation*. Honolulu: B. P. Bishop Museum, Special Publication 39. 154 p.

Vitousek, P. (2004). *Nutrient Cycling and Limitation: Hawaiʻi as a Model System*. Oxford and Princeton: Princeton University Press, Princeton Environmental Institute Series. 223 p.

Wirthmann, A. & Hueser, K. (1987). Vulkaninseln als Modelle tropischer Reliefgenese. Dargestellt am Beispiel von Hawaii, La Reunion und Mauritius. *Geographische Rundschau* 39(1): 22–31.

Chapter 6

Rebirth of ʻŌhiʻa Lehua After Collapse

Early into the investigation of the canopy collapse in all dieback types, we became aware of rebirth of ʻŌhiʻa lehua from new seedling establishment and its resilience as a species (Photo 6.1). Individuals and groups of the older generation (cohorts) were dying in the different dieback types, but new individuals also emerged in groups as new cohorts. The resilience of ʻŌhiʻa became obvious already on some of the less

Photo 6.1. Young ʻŌhiʻa seedling (three smaller plants) and a larger Kāwaʻu (*Ilex anomala*) seedling growing on a moss and liverwort covered Hāpuʻu tree fern trunk. Photo by H. J. Boehmer.

frequented jeep trails during the dieback phase of our investigation, where we observed vigorous seedlings in the center strip of such jeep tracks (Photo 6.2).

Moreover, we noted vegetative regrowth from fallen 'Ōhi'a trees in the bog-formation dieback (Photo 4.25 in Chapter 4, p. 106).

To recall from Chapter 5, the wetland and dryland types were identified structurally as replacement diebacks. The displacement dieback in the eutrophic soil phase also showed young 'Ōhi'a in the undergrowth, but there were rarely any saplings. In other words, there was an impediment for seedlings to become saplings due to competition from other species. The gap-formation and bog-formation diebacks in the regression

Photo 6.2. Young 'Ōhi'a saplings and Uluhe fern growing in the center of a jeep road running through dieback stands in the Saddle Road area on the island of Hawai'i.

phase were structurally identified as stand-reduction diebacks, meaning ʻŌhiʻa replacement was structurally reduced in height growth and spatial density. The only real die-off was associated with incipient stream formation, but the downstream banks offered rejuvenated soils that provided for subsequent seedling establishment or rebirth of ʻŌhiʻa lehua.

Thus our observations of ʻŌhiʻa rebirth suggested three kinds of future stand structures after canopy dieback:

(1) Replacement with ʻŌhiʻa lehua as the dominant canopy tree. This form of recovery relates to both the dryland dieback and the wetland dieback in the progression phase.

(2) Displacement of ʻŌhiʻa as future canopy tree by competition. This was originally seen primarily in the eutrophic climax phase and named displacement dieback, see Fig. 5.1 in Chapter 5 (p. 121).

(3) ʻŌhiʻa regrowth resulting in stand reduction relative to the parental forest. This relates to both the gap formation and bog formation diebacks in the regression phase of soil and landscape aging.

Replacement with ʻŌhiʻa lehua

The analysis of a large number of vegetation plots in healthy ʻŌhiʻa stands and plots in the dieback phase allowed for drafting conceptual ecological models of what may be happening in these forests. Fig. 6.1 is a size frequency model of healthy mature rainforests with closed ʻŌhiʻa tree canopies like those shown in several photos in Chapter 1.

The model in form of a size frequency graph is based on measuring tree height and diameter in a large number of mature stands with closed canopies. Such non-dieback or healthy stands which have a dense tree canopy and a shaded understory show typically a seedling cohort of numerous small individu-

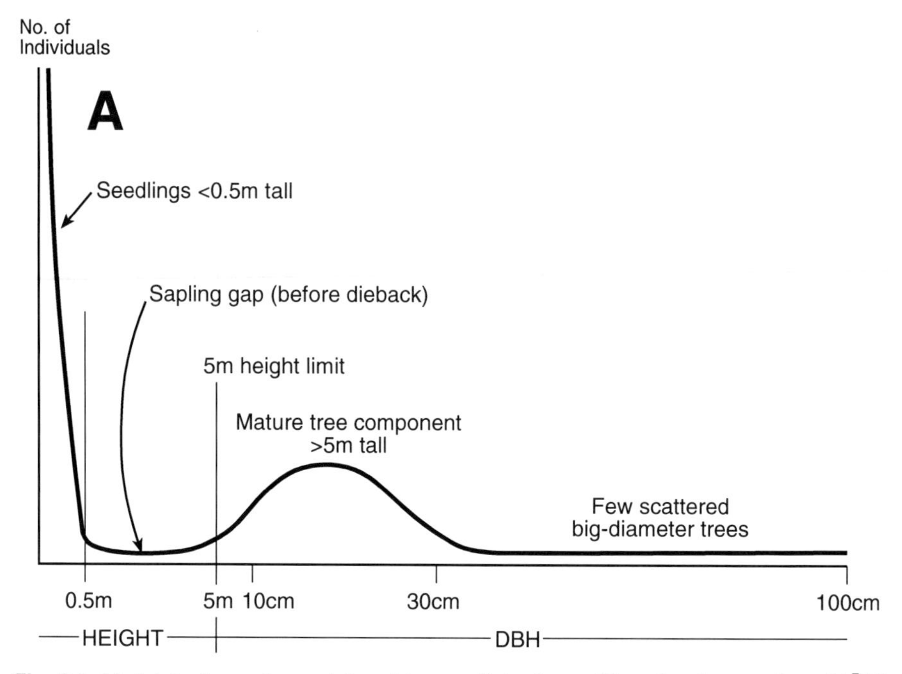

Fig. 6.1. Model A of a mature rainforest in non-dieback condition showing number of 'Ōhi'a lehua individuals over size classes, a size frequency graph.

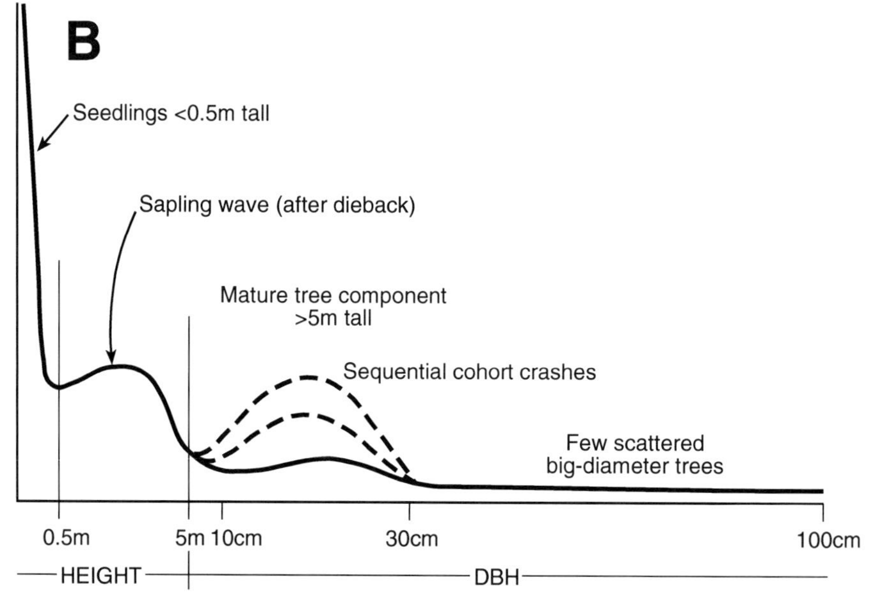

Fig. 6.2. Model B of a forest in dieback condition.

als under 10 cm (4 in) in height. Seedling numbers decrease sharply to a size group of half a meter (or one and a half feet) in height. There are hardly any ʻŌhiʻa individuals larger than 0.5 m in height until one encounters trees exceeding 5 m (15 ft). These are the smaller canopy trees from 2.5–10 cm (1–4 in) in diameter at breast height (DBH) that appeared to be suppressed by the larger diameter trees. Those from 20–30 cm (8–12 in) DBH form a normal or bell-shaped curve, thereby indicating a cohort in size structure. This curve is similar to that of an even-aged plantation forest. Note also that there are a few trees much larger in diameter. They occur usually as individuals across the broader territory of the same site. They seem to be survivors of a former canopy collapse, and may be twice as old as the prevailing canopy trees. This is an as yet unresolved mystery.

Fig. 6.2 depicts a model of a forest in dieback condition. Note that canopy dieback does not happen all at once. It manifested itself typically in sequential cohort crashes, an indication that the individuals in a canopy cohort were not all in the same vigor state or equally predisposed to dieback. Also a few large diameter ʻŌhiʻa trees remained unaffected as shown by the drawn out line to trees 100 cm (40 in) in DBH. An important ʻŌhiʻa replacement phase began concurrently with the dieback crashes. Seedlings began to grow into the sapling classes, and some time after the dieback we noted a “sapling wave” to develop in our permanent plots.

Permanent plots and dieback mapping

In 1976/77 we established 26 permanent plots, half in dieback and half in nearby non-dieback stands. All mature trees dead and alive were labeled and remeasured and ʻŌhiʻa regeneration was assessed; these plots were resampled at approximately five year intervals. Fig. 6.3 shows a map with 26 permanent plots that we followed up over a period of almost 30 years (Boehmer 2005).

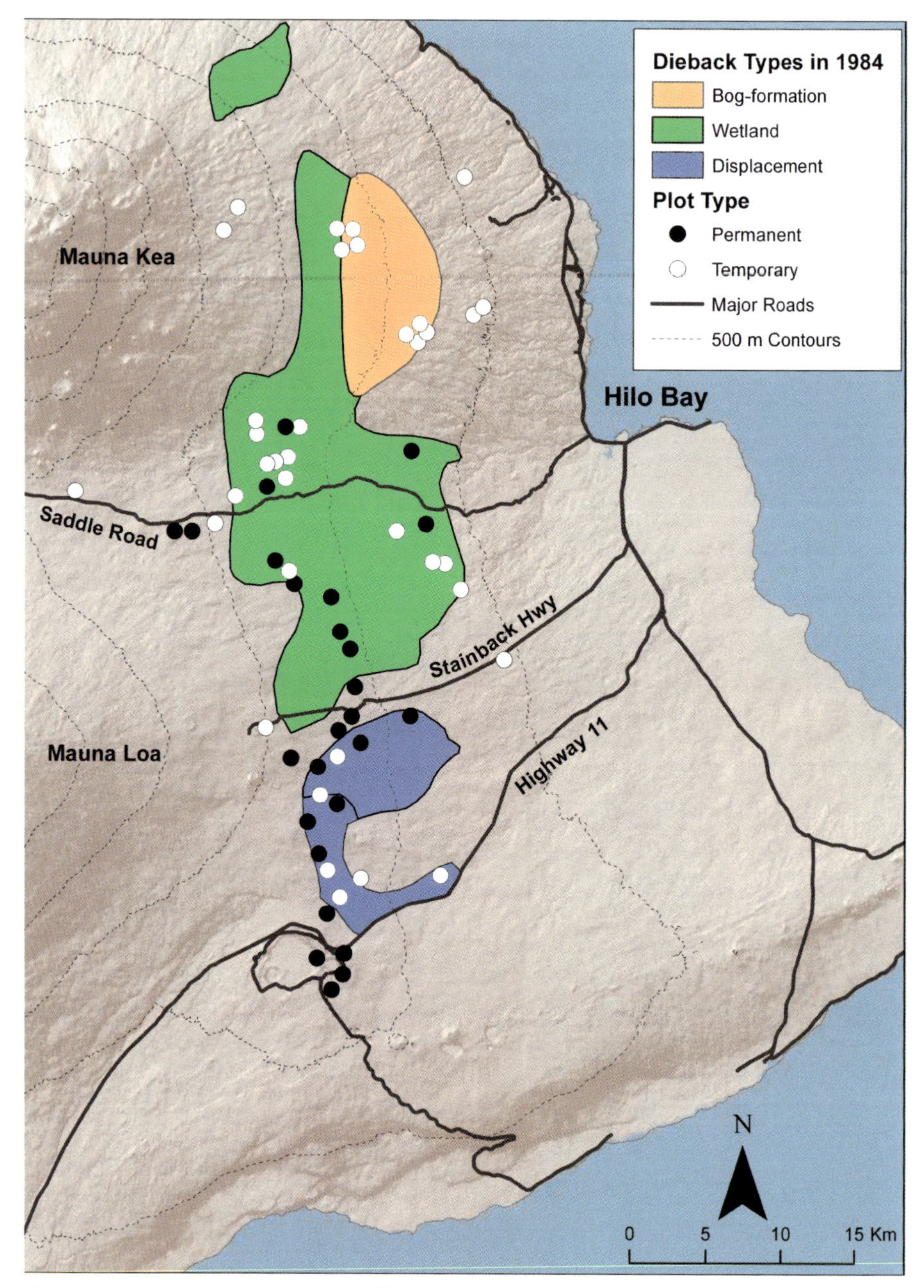

Fig. 6.3. Map showing distribution of ‘Ōhi‘a dieback on the windward slopes of Mauna Kea and Mauna Loa volcanoes. Circles indicate locations of the vegetation plots established to study dieback and non-dieback sites during the 1970s and 1980s; black circles are the 26 permanent plots that have been resampled since then.

Our permanent plots (26 out of a total of the 62 initially surveyed and analysed during our first decade of field research, from 1975–1985), were distributed from inside Hawaiʻi Volcanoes National Park northward across the Saddle Road as seen on Fig. 6.3.

The story of canopy rebirth

The following three figures tell the story of rebirth and recovery after dieback.

The ʻŌhiʻa seedlings in non-dieback plots remained in nominal numbers below 2000/ha, shown in Fig. 6.4 by the blue line from 1976–2003. Seedlings survive in the shade for a few years, and new ones are reborn steadily under the closed canopy of the parental generation. In contrast seedlings in dieback plots (red line) were rather numerous in the early stages of canopy breakdown, but decreased in the following years. The reason for the seedling decline in dieback stands becomes apparent in the next diagram, Fig. 6.5. Saplings were already more numerous in dieback plots in 1976 (red line). They developed into a wave in 1982 and remained numerically high. They declined in numbers after 10 years when individuals started to become trees forming the new canopy. Saplings in non-dieback plots (blue line) remained low in number through the same period from 1976–2003.

Following 1985, there was a steady increase of ʻŌhiʻa trees in most of the dieback plots. As seen on Fig. 6.6, complete numerical recovery was indicated by 2003 in the dryland and wetland dieback plots, which belong to the ʻŌhiʻa replacement category. Two plots in the ʻŌhiʻa displacement category behaved as expected, i.e., no replacement. Likewise two plots in the stand reduction category did not return to the original level of tree development (see Boehmer 2005). Another assessment is currently underway to assess the condition of former mapped

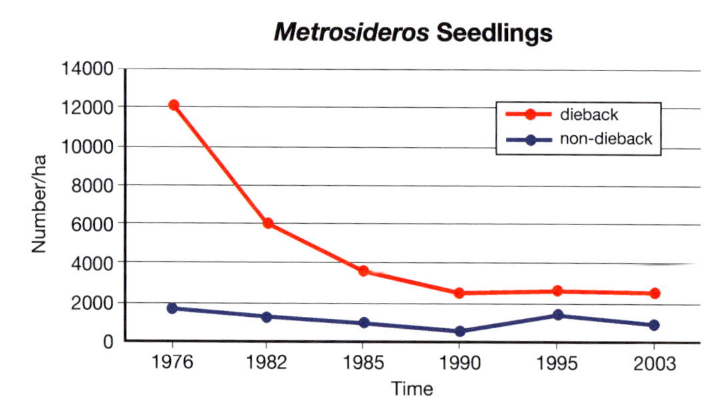

Fig. 6.4. 'Ōhi'a seedling numbers in dieback and non-dieback plots from 1976–2003. These were the small seedlings <10 cm = 4 in tall.

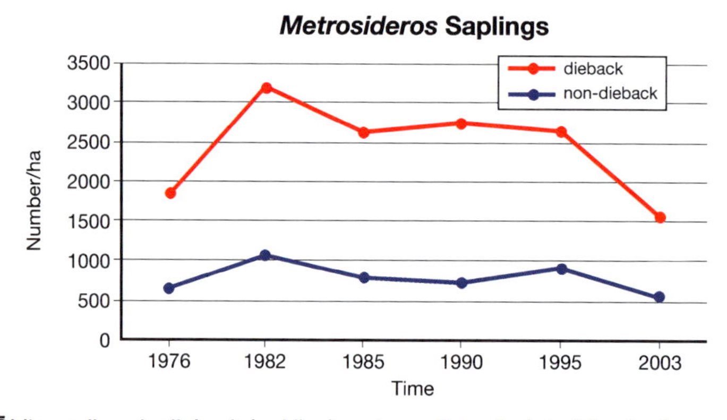

Fig. 6.5. 'Ōhi'a saplings in dieback (red line) and non-dieback plots (blue line).

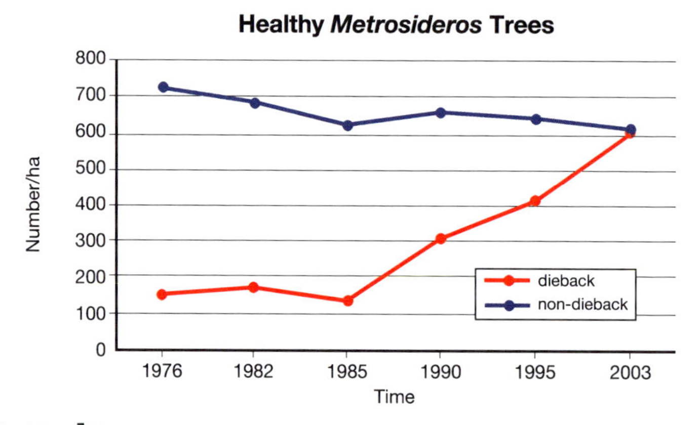

Fig. 6.6. Healthy 'Ōhi'a trees in dieback (red line) and non-dieback plots (blue line).

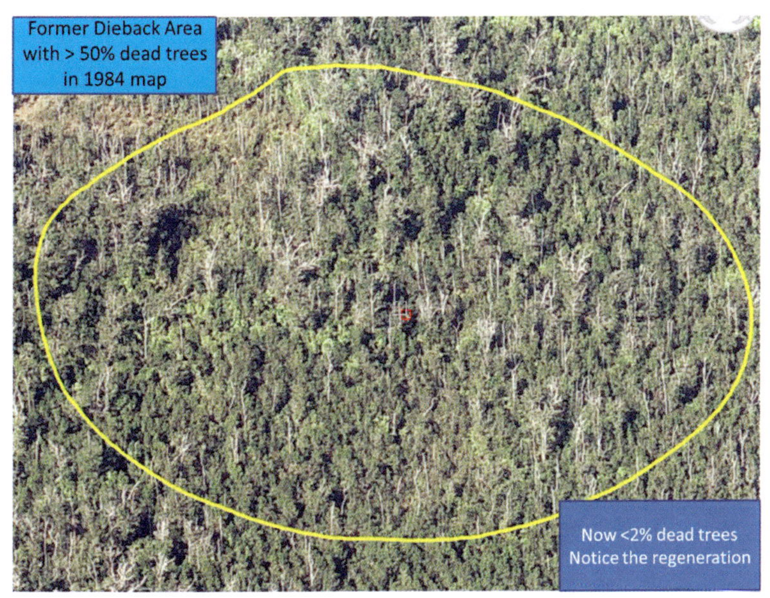

Photo 6.3. Oblique view of a former dieback forest on the windward slope of Mauna Kea showing canopy and understory status within a 100 m radius survey plot. Image courtesy of Pictometry International.

Photo 6.4. Another dieback assessment plot showing an extremely dense cohort of young ʻŌhiʻa trees growing into a new tree canopy in what was formerly mapped as an area with heavy canopy dieback. Image courtesy of Pictometry International.

dieback areas. This research, which is being conducted by Linda Mertelmeyer, a graduate student of H. Juergen Boehmer, is utilizing new aerial imagery taken by Pictometry International, to determine current conditions of the 'Ōhi'a tree canopy and regenerating cohorts. Although this study has not yet been completed, preliminary results suggest that many areas that were formerly mapped as having moderate to severe dieback, now have a relatively healthy 'Ōhi'a tree canopy as well as a well developed and vigorous 'Ōhi'a sapling cohort (Photo 6.3 and 6.4).

Displacement of 'Ōhi'a

Prior to the more recent biological invasion syndrome, displacement of 'Ōhi'a appears only to be a transient phenomenon in the eutrophic climax phase of primary succession. The undergrowth tree ferns did suppress the full recovery and resurgence of the next 'Ōhi'a canopy after dieback in this phase of nutrient-rich soil substrate development. Moreover, 'Ōhi'a itself had become evolutionary adjusted and displaced earlier varieties with those more adapted to changing soils and landscapes.

Today, the situation has changed. Human caused introductions of so many new species has facilitated the escape of species that are alien to the native and natural composition of the Hawaiian rainforest. This more recent anthropogenic interference has brought about a new competitive environment for 'Ōhi'a lehua. This change is recognized as the principal cause of fragility in the Hawaiian rainforest and in most other island ecosystems worldwide. Specifics of this new fragility will be discussed in the next chapter.

'Ōhi'a regrowth and stand reduction

Not all 'Ōhi'a dieback stands recovered into new forests. This has become clear from several decades of observations made in the historic dieback territory of the island of Maui. The reason became obvious because here the forest landscape is gradually changing into a boggy landscape with occasional stream formation. The same was observed on Mauna Kea during our research there. Where the boggy habitats are drained by stream formation, ʻŌhiʻa is coming back on slopes cut from streams. Thus ʻŌhiʻa is tenaciously holding on to its territory in form of scattered trees or small stands invading new slopes and in patch communities recovering on dead trees and woody peat among the sun-loving Uluhe fern. This fern often forms contiguous mats and thereby the matrix of such reduced ʻŌhiʻa shrub forests.

Certainly, ʻŌhiʻa is not a swamp tree. Swamp trees are specialists which occur naturally for example in the Solomon Islands. But ʻŌhiʻa is remarkably resilient under adverse conditions. This aspect of resilience will be discussed in the next chapter with specific examples.

Suggested Readings

Akashi, Y. & Mueller-Dombois, D. (1995). A landscape perspective of the Hawaiian rainforest dieback. *Journal of Vegetation Science* 6: 449–464.

Boehmer, H. J. (2005). Dynamik und Invasibilitaet des montanen Regenwaldes auf der Insel Hawaii [Dynamics and Invasibility of Hawaiiʻs Montane Rainforest]. Habilitation Thesis, Department of Ecology and Ecosystem Managment, Technical University of Munich, Germany. 232 p. Includes six appendices.

Boehmer, H. J., Wagner, H. H., Jacobi, J. D., Gerrish, G. C. & Mueller-Dombois, D. (2013). Rebuilding after Collapse: Evidence for long-term cohort dynamics in a monodominant tropical rainforest. *Journal of Vegetation Science*. DOI: 10.1111/jvs.12000.

Burton, P. J. & Mueller-Dombois, D. (1984). Response of *Metrosideros polymorpha* seedlings to experimental canopy opening. *Ecology* 65(3): 779–797.

Hodges, C. S., Adee, K. T., Stein, J. D., Wood, H. B. & Doty, R. D. (1986). *Decline of ʻŌhiʻa (Metrosideros polymorpha) in Hawaii: A review*. Berkeley: US Department of Agriculture, Forest Service, Pacific Southwest Forest and Range Experiment Station. General Technical Report PSW-86. 22 p.

Holt, R. A. (1983). The Maui forest trouble: A literature review and proposal for research. *Hawaiʻi Bot. Sci. Paper* 42: 1–67. URL: www.botany.hawaii.edu/pabitra.

Jacobi, J. D. (1983). *Metrosideros* dieback in Hawaiʻi: a comparison of adjacent dieback and non-dieback rain forest stands. *New Zealand Journal of Ecology* 6: 79–97.

Jacobi, J. D., Gerrish, G. & Mueller-Dombois, D. (1983). 'Ōhi'a dieback in Hawai'i: Vegetation changes in permanent plots. *Pacific Science* 37(4): 327–337.

Jacobi, J. D., Gerrish, G., Mueller-Dombois, D. & Whiteaker, L. (1988). Stand-level dieback and *Metrosideros* regeneration in the montane rain forest of Hawaii. *GeoJournal* 17(2): 193–200.

Mueller-Dombois, D. (1987). Natural dieback in forests. *BioScience* 37(8): 575–583.

Santiago, L. S., Goldstein, G., Meinzer, F. C., Fownes, J. H. & Mueller-Dombois, D. (2000). Transpiration and forest structure in relation to soil waterlogging in a Hawaiian montane cloud forest. *Tree Physiology* 20(10): 673–681.

Selling, O. H. (1948). Studies in Hawaiian pollen statistics. In *Part III: On the Late Quaternary History of the Hawaiian Vegetation.* Honolulu: B. P. Bishop Museum, Special Publication 39. 154 p.

Chapter 7

Fragility vs. Resilience in the Hawaiian Rainforest

The Hawaiian rainforest can be characterized by both fragility and resilience.

Fragility can be defined as something "easily broken." The term fragility brings us back to the debate at the start of the Hawai'i IBP (International Biological Program in Hawai'i) in August 1970 (see Chapter 3, p. 79), when the concept of fragility was taken for granted as applying to island ecosystems as compared to continental ecosystems. As a result, the IBP research focus was first directed to the study of intact native island ecosystems, which included the Mauna Loa transect ecosystems and the Kīlauea rainforest. The 'Ōhi'a dieback/decline problem was considered "applied research" to be left to the US Forest Service researchers. The IBP thinking was that the concept of stability did not apply to islands but was more applicable to continental ecosystems. The greater canopy species diversity of continental tropical ecosystems, for example, gave support to the stability concept. That concept implies "strength with endurance," the sort of picture one gets from visiting a mature rainforest. In contrast, the concept of resilience implies to "comeback" after a disturbance. In Chapter 6, we used the term "rebirth," which certainly is an aspect of resilience.

Examples of fragility

Fragility in the Hawaiian rainforest can be observed in a number of situations, but all have to do with the invasion of alien organisms, those organisms accidentally or willfully introduced by humans. Fragility situations for 'Ōhi'a lehua trees can be classified into four groups, 'Ōhi'a life cycle disrupters, 'Ōhi'a canopy displacers, 'Ōhi'a killer trees, and 'Ōhi'a disease threats.

'Ōhi'a life cycle disrupters

The 'Ōhi'a rainforest in the village of Volcano is often invaded by a vigorous alien shrub on abandoned properties, the lasiandra or princess flower (*Tibouchina urvilleana*). These shrubs start to grow in disturbed areas outside the forest but can form dense thickets beneath the existing 'Ōhi'a tree canopy (Photo 7.1).

Dryland dieback has occurred in several places around the village. Wherever *Tibouchina* thickets are found growing un-

Photo 7.1. *Tibouchina urvilleana* shrub invasion on an abandoned lot in Volcano village. This ornamental shrub is an absolute life-cycle disrupter for the 'Ōhi'a canopy trees because it develops into long-lived thickets in the understory that leave no room for 'Ōhi'a seedlings to become established and grow to maturity. Inset photo by Forest and Kim Starr.

der dying ʻŌhiʻa canopy trees, there will be no replacement or rebirth with ʻŌhiʻa. If this alien undergrowth is left in place, the dying ʻŌhiʻa cohort will surely be the last native tree generation in that location.

Another species that prevents ʻŌhiʻa rebirth under a mature native rainforest canopy is the Kāhili ginger (*Hedychium gardnerianum*) (Photo 7.2). This is likewise a beautifully flowering introduced ornamental garden escapee (as is the princess flower). Its native land is the Himalayas. It has the capacity to grow in open sunlight as well as in the shade. In Hawaiʻi Volcanoes National Park it has become locally a dense undergrowth invader that blocks the light at the ground level and leaves no room for native tree or shrub seedlings to become established. The only tree that can penetrate the dense ginger cover is the alien strawberry guava (*Psidium cattleianum*).

Photo 7.2. Kāhili ginger (*Hedychium gardnerianum*) in the undergrowth of the 200 year old rainforest in Hawaiʻi Volcanoes National Park. This plant was introduced as a garden ornamental but is now a serious invasive species in Hawaiʻi Volcanoes National Park. It develops foliage and root tubers that become so dense that it completely restricts the regeneration of ʻŌhiʻa seedlings following canopy dieback.

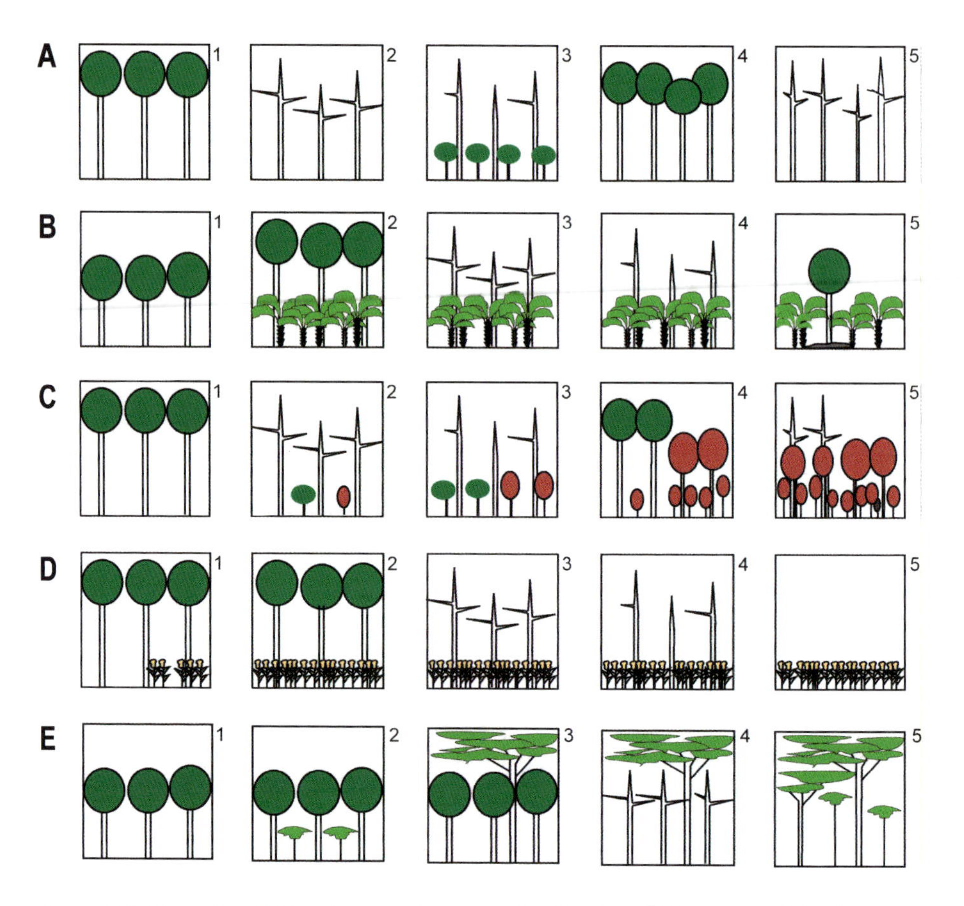

Figure 7.1. Example of different succession scenarios following 'Ōhi'a canopy dieback.

A) 'Ōhi'a replacement dieback (natural, by auto-succession).

B) 'Ōhi'a displacement dieback (natural, with tree ferns forming canopy).

C) 'Ōhi'a displacement dieback (canopy displacement with alien *Psidium*).

D) 'Ōhi'a life-cycle disruption by dense and persistent weed patch (Kāhili ginger).

E) 'Ōhi'a killer tree dieback (by alien albizia imposing light starvation).

(after Boehmer & Niemand 2009).

'Ōhi'a canopy displacers

Fragility can also be caused by alien tree invaders (Fig. 7.1). First among these is the above mentioned strawberry guava, also named Waiawī (Photo 7.3 and 7.4). This tree has fruits that taste almost like strawberries. It was introduced originally for this reason from tropical America. It is a favored food of feral pigs (*Sus scrofa*), who promote thicket development of this tree through their rooting activity. Moreover, the fruit is consumed by alien birds, which distribute the seeds widely. These are two distribution advantages competing with the wind dispersal of 'Ōhi'a seeds.

Another, at least partial displacer, particularly in Hawai'i Volcanoes National Park, is the Faya tree (*Morella faya,* formerly named *Myrica faya*) (Photo 7.5). Native to the Azores, Madeira and Canary Islands, it was introduced to the island of Hawai'i

Photo 7.3. A juvenile 'Ōhi'a rainforest being invaded by Strawberry guava (*Psidium cattleianum*), a canopy displacer of 'Ōhi'a, often forming dense thickets.

Photo 7.4 (left). Close-up of leaves and fruits of Strawberry guava. This small tree can form extremely dense thickets under the canopy of ʻŌhiʻa forests, particularly in the lowland wet forest habitats. Photo by Forest and Kim Starr.

Photo 7.5. Invasion of the introduced Faya tree (*Morella faya*) into an open ʻŌhiʻa forest along the Chain of Craters Road in Hawaiʻi Volcanoes National Park.

in 1926 at several sites along the windward Hāmākua coast. It appeared suddenly in Hawai'i Volcanoes National Park in about 1968, when it was probably brought in by alien birds, most likely the Japanese white eye (*Zosterops japonica*).

A major advantage that Faya trees have is their association with a nitrogen fixing Actinomycete (*Frankia*), which provides for accelerated growth in low nitrogen substrates such as volcanic cinder. Faya trees also have the distribution advantage of their seed imbedded in juicy fruits that are spread around by the White eye, Mynah birds (*Acridotheres tristis*), and perhaps the native Nēnē geese (*Branta sandvicensis*).

'Ōhi'a "killer" trees

A number of introduced tree species can overtop 'Ōhi'a trees, thereby shading them out by light starvation which ultimately can lead to the death of the trees. A classic example is the Albizia tree (*Falcataria moluccana*) (Photo 7.6, 7.7, and 7.8). It is widespread across the modified lowland rainforest on several of the Hawaiian Islands. On the island of Hawai'i it has also invaded lava flows that were initially occupied by young cohorts of 'Ōhi'a. Albizia, being a legume tree, comes equipped with its own symbiotic nitrogen fixing bacteria (*Rhizobium* spp.), which gives it a great advantage on young volcanic substrates that are notoriously low in the supply of nitrogen. Thus albizia overgrows 'Ōhi'a rapidly on such substrates.

Another tree with similar abilities is the introduced Ironwood tree (*Casuarina equisetifolia*), which has become widely established on volcanic cinder deposits in the Kapoho area of Hawai'i island (Photo 7.9). Both these tree species have pioneer qualities as does 'Ōhi'a, but they can exclude the native colonizer on pioneer surfaces in the warm lowland rainforest environment.

A most threatening ornamental tree introduction from tropical America is *Miconia calvescens*, a member of the melas-

Photo 7.6. Albizia (*Falcataria moluccana*) thriving on 1977 ʻaʻā flow with an ʻŌhiʻa seedling cohort starting to develop in the foreground. If albizia continues to spread, the ʻŌhiʻa cohort will be overtopped and eventually killed. The photo was taken in the lowland rainforest habitat near Kalapana Village on the island of Hawaiʻi in 1994, 17 years after the lava flow had cooled.

Photo 7.7. A juvenile ʻŌhiʻa cohort growing on a 1955 lava flow adjacent to a dense albizia stand that is spreading across the landscape. Photo taken in 1984.

Photo 7.8. Juvenile ʻŌhiʻa stand being overtopped by Albizia, a "killer" tree.

Photo 7.9. Ironwood (*Casuarina equisetifolia*) trees outgrowing ʻŌhiʻa on a cinder deposit in 1984 near Kapoho on the island of Hawaiʻi. Ironwood is another killer tree in the warm coastal lowland rainforest environment.

Photo 7.10. *Miconia calvescens* is another potential killer tree here shown by Jean-Yves Meyer in Tahiti. Miconia is currently spreading across much of the wet lowland habitats on the island of Hawaiʻi and has the potential to also form a very dense subcanopy shade layer, significantly restricting the potential for regeneration of native plant species and natural succession of the ʻŌhiʻa forest.

tome family. This tree is both shade and light tolerant, which means it can get established and grow under a closed canopy and then push its crown through an established native tree canopy (Photo 7.10). This has occurred in Tahiti, where this miconia has had far reaching effects on displacing native forest (Jean-Yves Meyer 1996).

Because of its aggressively invasive behavior in Tahiti, miconia has been declared a noxious species in Hawai'i. It was originally planted as an ornamental on private properties in the Hilo area and has now spread both around the coast as well upslope. The tree is easily spotted because of its huge (up to 1 m = 3 feet long) dark-green velvety leaves with maroon-purplish undersides. It also produces abundant seeds in small purple berries that are easily distributed by birds.

Many other shade adapted (late-successional) and tall-growing tree species have been introduced and planted in the Hawaiian islands, making it impossible for 'Ōhi'a trees to survive in some of its former rainforest habitat.

'Ōhi'a disease threats

Since the native 'Ōhi'a forest evolved in isolation for over 5 million years, threats of new killer diseases are always to be expected. The term killer disease applies to diseases for which the native plants are not adapted, lacking adequate mechanisms for defense. No doubt this was the expectation which led to the early conviction by the US foresters and pathologists to search for a killer disease when the 'Ōhi'a canopy dieback was first discovered. It must be re-emphasized that the negative finding of this search for a killer disease causing the widespread 'Ōhi'a dieback on the windward side of the island of Hawai'i in the 1960s and 1970s was a most important result.

Recently, however, a new rust fungus has appeared in Hawai'i. Its name is *Puccinia psidii*, native to Brazil. The name implies association with species of the genus *Psidium*, which includes the above mentioned Strawberry guava. But it also includes ʻŌhiʻa, since both species are members of the myrtle or eucalypt family. Since its discovery in 2005, this *Psidium* rust has spread widely in Hawaiʻi by means of light-weight spores. It has been detected on some ʻŌhiʻa trees first in a nursery on Oʻahu, then on Maui and Hawaiʻi, but always with inconsequential injury. However, the *Puccinia* rust has radically infected and killed the introduced Rose apple tree (*Syzygium jambos*) in a number of places in the islands (Photo 7.11), as well as a rare native tree species (*Eugenia koolauensis*), also closely related to ʻŌhiʻa.

So far this new rust fungus has seriously impacted only a few species in Hawaiʻi, but in its native Brazil it has infected a wider host range, from common guava (*Psidium guajava*) to many *Eucalyptus* species that had been planted in timber planta-

Photo 7.11. Rose apple (*Syzygium jambos*) tree thicket in the process of dying after being infected by the introduced rust *Puccinia psidii*.

tions. This led to the suspicion that different genetic strains are involved. A study of genetic strains in *Psidium* rust is currently underway in Brazil. Also extremely serious precautions are now taken by the Hawai'i Department of Agriculture to restrict importation of members of the myrtle family, which includes Eucalypts and many other species related to 'Ōhi'a.

Examples of resilience

Resilience, the strength to survive under adverse conditions and to come back restored from a disturbance, has been shown for 'Ōhi'a in many situations. Certainly, the 'Ōhi'a replacement dieback (Chapter 6) is a prime example.

But here we look at situations involving human made landscape changes wherein 'Ōhi'a has demonstrated resilience.

Photo 7.12. A small group of yellow flowering 'Ōhi'a trees surrounded by native 'Ōkupukupu (*Nephrolepis exaltata*) fern that remain in a heavily disturbed road side near Honoka'a on the windward coast of the island of Hawai'i.

'Ōhi'a trees in pastures

'Ōhi'a trees show remarkable tenacity in being able to remain in sites that have been heavily disturbed. One of the best examples of this can be seen in former rainforest areas that have been converted to cattle pasture (Photos 7.12 through 7.18). Although the mature 'Ōhi'a trees appear able to live as individuals or small stands within pastures, the grazing and trampling by cattle suppresses regeneration of 'Ōhi'a and eventually the trees die without being replaced (Photo 7.19).

Photo 7.13 (left). A small patch of remnant 'Ōhi'a rainforest in a pasture along Highway 19 near Honoka'a on the island of Hawai'i. Notice that Hāpu'u tree ferns are still well represented in this stand.

Photo 7.14. A lone 'Ōhi'a tree with two trunks still survives in a wide open pasture below Highway 19 near Honoka'a on the island of Hawai'i.

Photo 7.15. A row of ʻŌhiʻa trees growing on a road cut along Highway 19 on the island of Hawaiʻi. The trees must have regenerated in this area naturally after the mineral soil was exposed during road construction.

Photo 7.16. An open mature ʻŌhiʻa stand with broad crowns growing in a pasture above Highway 19 near Honokaʻa on the island of Hawaiʻi.

Photo 7.17. A closed mature 'Ōhi'a stand in the same pasture. This stand demonstrates the resilience of 'Ōhi'a in a heavily modified landscape.

Photo 7.18. A tall 'Ōhi'a stand remaining in lowland pasture below the 1984 clear-cut logging area in the rainforest above Kalapana on the island of Hawai'i.

Photo 7.19. Tall senescing ʻŌhiʻa trees in a pasture on the Kona side of the Island of Hawaiʻi.

ʻŌhiʻa rebirth after clearcut logging

The inherent pioneering capacity of ʻŌhiʻa allows it to quickly recolonize other disturbed sites, such as landslides or clear cut forests. For example, in 1984 a section of dense lowland ʻŌhiʻa forest in the Puna District on the island of Hawaiʻi was clear cut to provide wood chips to feed a biofuel generator. While there was considerable controversy over the utility versus impacts of this project, it provided an opportunity to study the regeneration of ʻŌhiʻa, as well as other associated native and alien plant species, once the project ended (Photo 7.20). Dennis Grossman, who was a graduate student at that time, initiated a study of plant succession in this area. He established a set of monitoring plots and seed traps to document changes in plant species composition and vegetation structure over time.

Within 10 years of the original clear-cut disturbance, a mix of native and alien plants were found to be colonizing this area, including many young 'Ōhi'a saplings (Photos 7.21 to 7.25). By 2005, it was clear that many of the 'Ōhi'a saplings were growing into larger trees and would eventually form a closed canopy again in some of the former clear-cut areas. However, many invasive alien grasses, shrubs, and ferns were still found to be very abundant in the understory and are likely to impede full succession of the vegetation back to a native lowland rainforest (Photo 7.26 to 7.29).

Photo 7.20. Tall (30 m=100 ft) 'Ōhi'a stand left standing at the upper margin of the 1984 clear-cut area in the lowland rainforest above Kalapana on the island of Hawai'i in 1987.

Photo 7.21. View of the Kalapana cut-over forest area 10 years after all of the trees were removed for wood chips. The few remaining ʻŌhiʻa trees grew new branches along their trunks where they had been injured during the logging operation. The grass in the foreground is Broomsedge (*Andropogon virginicus*), and the shrubs are mostly Sourbush (*Pluchea carolinensis*), both of which are early invaders in secondary succession in this type of habitat. The small tree in the center of the photo with broad, ovate, pale-green leaves is *Melochia umbellata*, an introduced member of the cacao family, and the white flowers are from the introduced Bamboo orchid (*Arundina graminifolia*).

Photo 7.22. One of the seed traps installed on the site by Dennis Grossman, who studied early secondary succession of this clear-cut ʻŌhiʻa forest for his PhD dissertation in 1990 (see p. 228).

Photo 7.23. To the left of the seed trap is a young 'Ōhi'a sapling growing among the native 'Ōkupukupu (*Nephrolepis* sp.) ferns and introduced weeds.

Photo 7.24. By 1994 two 'Ōhi'a saplings were found pushing through the Sourbush understory. The red leaves on the 'Ōhi'a are typical of the young growth for this glabrous (smooth) leaved variety of 'Ōhi'a.

Photo 7.25. View of the Kalapana cut-over site in 1995 with Kanehiro Kitayama standing on the side of a former logging road.

Photo 7.26. Annette Mueller-Dombois standing beside a pink flowering ʻŌhiʻa sapling growing among a mix of native sedges and shrubs, as well as many introduced species in the former cut-over forest near Kalapana on the island of Hawaiʻi. By this time the native species, and ʻŌhiʻa in particular, appear to be successfully recolonizing the site and suppressing many of the introduced invaders.

Photo 7.27. View in 2005 of the logging road shown in Photo 7.25. The 'Ōhi'a trees have grown substantially and appear to be in the process of forming a more dense tree canopy.

Photo 7.28. A new 'Ōhi'a cohort stand near the entrance gate to the former clear-cut area.

Photo 7.29. By 2005, ʻŌhiʻa trees are now quite tall and are even growing through the burned-out loggers' bus in the Kalapana cut-over forest site.

ʻŌhiʻa dense stand redevelopment after removal of tree ferns

Although ʻŌhiʻa forests can show strong resilience to habitat disturbance, this may be compromised by the presence of highly invasive species. For example, in an area immediately adjacent to the Ōlaʻa Tract section of Hawaiʻi Volcanoes National Park, the removal of tree ferns from the understory nearly 100 years ago resulted in a cohort regeneration of ʻŌhiʻa which now forms a distinct, nearly closed canopy, as opposed to the open ʻŌhiʻa canopy in the undisturbed section of the National Park (Photo 7.30). Fortunately, there were no strong invaders present at the time of initial disturbance and ʻŌhiʻa seedlings were able to respond with rapid growth. However, if this type of disturbance were to occur in the same area today, the opened understory would likely be quickly invaded by non-native species such as Kāhili ginger, *Tibouchina*, or *Faya* tree, and there would not be a replacement cohort of ʻŌhiʻa.

Photo 7.30. Two adjacent properties at Wright Road in Volcano Village: at left 'Ōhi'a displacement dieback has exposed the tree fern undergrowth; at right a dense 'Ōhi'a rainforest has redeveloped after the tree ferns were removed. The same 'Ōhi'a forest regeneration can be expected wherever the introduced 'Ōhi'a life-cycle disrupters, such as *Tlbouchina* or the Kāhili ginger are kept from invading disturbed native forest sites. This image was taken by Pictometry International and obtained through their Pictometry Online image viewer.

Ōhi'a populations surviving in challenging situations

Here are three photos showing 'Ōhi'a populations in extremely wet rainforest environments. Trees have assumed different life forms, a shrub-form on steep wind-swept slope (Photo 7.31) and a dwarf-form in upland (Photo 7.32) and lowland (Photo 7.33) bogs.

The question of resilience in these challenging situations is: Are these life forms genetic adaptations or just acclimations i.e., expressions of phenotypic plasticity.

The answer can be experimentally tested by planting seed of these populations in our **Common Garden**, and observing their development over time. This common garden is a testing ground for 'Ōhi'a varieties or ecotypes located at the Agricultural Experiment Station in Volcano Village.

COMMON GARDEN

This is a plot of ground or greenhouse bench where different types of plants can be grown side-by-side under carefully managed environmental conditions. A common garden was established in the Hawai'i Volcanoes National Park greenhouse facility by Lani Stemmermann to study altitudinal and successional ecotypes of 'Ōhi'a under identical soil moisture conditions. This common garden is still being used as a resource for genetic research on 'Ōhi'a.

Conclusions

Fragility, and the opposite term resilience, are general attributes to characterize ecosystems.

Island ecosystems are often characterized as being fragile. Their fragility relates to having evolved in isolation without the rigor of competition from some stressors (e.g., ungulates) that were not naturally able to disperse and become established in

Photo 7.31. Native rainforest scrub community dominated by stunted 'Ōhi'a shrubs on a wind swept slope above Pelekunu Valley on the island of Moloka'i. Photo by J. Jacobi.

Photo 7.32 (left). Dwarf 'Ōhi'a population growing in the Pepe'ōpae bog in upland rainforest area on the island of Moloka'i. The 'Ōhi'a pollen diagram in Chapter 5, Fig. 5.3 was collected from this bog.

Photo 7.33. Another dwarf 'Ōhi'a population growing in Kanaele (Wahiawā) bog on the island of Kaua'i at approximately 490 m (1600 ft). Note the red flowers of 10 cm tall individuals in the foreground.

remote archipelagos like the Hawaiian Islands. In pre-human times the conditions that result in apparent fragility of the Hawaiian ecosystems were not present (i.e., invasive species) and natural succession, including canopy dieback, was able to proceed to predictable stages. Competition for survival is expected to be more severe in continental regions, which harbor a wider spectrum of species with enhanced functional attributes. For the same reasons, alien invaders are often superior competitors. But this is not always the case.

Gap-formation dieback in the regression phase of landscape aging in Hawaiʻi can be considered an aspect of natural fragility. This form of dieback often leads to larger open patches that become overgrown with the native Uluhe fern (*Dicranopteris linearis*). In some locations, particularly on the older islands, gap-formation has resulted in ʻŌhiʻa fern savannas.

However, the mat-forming Uluhe fern is an important native rainforest component in the early and late stages of primary succession. It forms a major obstacle for the invasion of alien species in the regression phase of primary succession. In the early stages of primary succession, Uluhe fern assists in the formation of ʻŌhiʻa cohort development by sealing off further undergrowth succession in the juvenile ʻŌhiʻa rainforest. When the juvenile forest becomes mature, the Hāpuʻu tree ferns create dense shade which will lead to the displacement of the Uluhe fern upon closure of the ʻŌhiʻa tree canopy. In the late regression phase, the process is often reversed.

The natural rainforest actually displays more native resilience than fragility. But what about the bog-formation dieback, which results in loss of forest habitat, in stand reduction and dwarfing of ʻŌhiʻa? Here, fragility of the native rainforest becomes rather obvious, but only from the perspective of loss of the forest structure. The resulting native-dominated bogs have their own unique suite of biodiversity. This fragility led Harold

Lyon to his concept of promoting the introduction of the missing climax tree component.

However, with the planting of introduced Swamp mahogany and paperbark trees in soggy habitats, the landscape change from forest habitat to bogs and streams can only be slowed down. The change, we know now, is due to geomorphologic aging of the volcanic shield on the lower east slope of Haleakalā mountain. This process supercedes all efforts of forest management.

In contrast, active forest management can do a lot to reduce the impact of the alien invaders. Pastures can be restored to native forest by scarifying the grass mat with a bulldozer, then dropping tree fern trunks on the grass to allow ʻŌhiʻa and native ferns to reestablish themselves. Other invaders that interfere with the native forest growth cycle may be thinned or removed where feasible. Very important is the selective removal of killer tress such as albizia, that kill ʻŌhiʻa by light starvation.

It will not be possible to feasibly restore invaded native ʻŌhiʻa rainforests to their original state in our present environment, which has been changed so much through human influences. Many of our native ecosystems may be composed of a mix of native and alien species in the future. However, the resilience of the ʻŌhiʻa tree will certainly be a boost in preserving what is left of the native heritage associated with the Hawaiian rainforest ecosystem. ʻŌhiʻa is still the dominant canopy tree in most Hawaiian ecosystems (Photo 7.34). It can be called the "resilient protector" of what is left of the natural heritage in the native Hawaiian rainforest.

Photo 7.34. Mature 'Ōhi'a rainforest near Pu'u Ali'i on the island of Moloka'i. Photo by J. Jacobi.

Suggested Readings

Anderson, C. R. & Uchida, J. Y. (2008). *Disease Index for the Rust Puccinia psidii on Rose Apple in Hawai'i.* Honolulu: University of Hawai'i at Mānoa, College of Tropical Agriculture and Human Resources, CTAHR Plant Disease publication PD-37. 12 p.

Boehmer, H. J. & Niemand, C. (2009). Die neue Vegetationsdynamik pazifischer Wälder [The new dynamics of Pacific forests]: Wie Klimaextreme und biologische Invasionen Inselökosysteme verändern [How climatic anomalies and biological invasions change island ecosystems]. *Geographische Rundschau* 61(4): 32–37.

Huenneke, L. F. & Vitousek, P. M. (1990). Seedling and clonal recruitment of the invasive tree *Psidium cattleianum*: implications for management of native Hawaiian forests. *Biological Conservation* 53: 199–211.

Loope, L. L. & Uchida, J. (2011). The challenge of retarding erosion of island biodiversity through phytosanitary measures: an update on the case of *Puccinia psidii* in Hawai'i. *Pacific Science* 66(2): 127–139.

Loope, L. L. & Mueller-Dombois, D. (1989). Characteristics of invaded islands, with special reference to Hawaii. In *Biological Invasions: a Global Perspective*, ed. by J. A. Drake, et al. Chinchester, NY: Scientific Committee on the Problems of the Environment (SCOPE), International Council of Scientific Union, John Wiley & Sons, Ltd. pp. 257–280.

Meyer, J.-Y. (1996). Status of *Miconia calvescense* (Melastomaceae), a dominant invasive tree in the Society Islands (French Polynesia). *Pacific Science* 50(1): 66–76.

Minden, V., Jacobi, J.D., Porembski, S. & Boehmer, H. J. (2010). Effects of invasive alien kahili ginger (*Hedychium gardnerianum*) on native plant species regeneration in a Hawaiian rainforest. *Applied Vegetation Science* 13: 5–14.

Mueller-Dombois, D. & Loope, L. L. (1990). Some unique ecological aspects of oceanic island ecosystems. *Monogr. Syst. Bot. Missouri Bot. Garden* 32: 21–27.

Stone, C. P., Smith, C. W. & Tunison, J. T. (1992). *Alien Plant Invasions in Native Ecosystems of Hawaii: Management and Research.* Honolulu: University of Hawai'i, Cooperative National Park Resources Studies Unit. 903 p.

Smith, C. W. (1985). Impact of alien plants on Hawaii's native biota. In *Hawaii's Terrestrial Ecosystems: Preservation and Managemement*, ed. by C. P. Stone & J. M. Scott. Honolulu: University of Hawai'i, Cooperative National Park Resources Studies Unit. pp. 180–250.

Uchida, J. Y., Anderson, R. C., Kadooka, C. Y., de la Rosa, A. M. & Coles, C. (2008). *Disease Index for the Rust* Puccinia psidii *on 'Ōhi'a* (Metrosideros polymorpha) *in Hawai'i.* Honolulu: University of Hawai'i at Mānoa, College of Tropical Agriculture and Human Resources, CTAHR Plant Disease publication PD-38. 16 p.

Vitousek, P. M., Walker, L. R., Whiteaker, L. D., Mueller-Dombois, D. & Matson, P. A. (1987). Biological invasion of *Myrica faya* alters ecosystem development in Hawai'i. *Science* 138: 802–804.

Chapter 8

Global Outreach of the ʻŌhiʻa Dieback Story

Is abiotic forest dieback unique in Hawaiʻi?

The answer clearly is no. Abiotic forest dieback is in fact quite common, and it cannot only be blamed on global warming or climate change, which is a newly emerging emphasis in forest decline research (*Responses of Northern U.S. Forests to Environmental Change*, Springer Ecological Studies 139, 2000).

Research in 1981 was aimed at studying the vegetation of the southwestern Pacific islands. Dieback in forests with only one or two leading canopy species was seen in New Caledonia, Papua New Guinea and in several places within New Zealand. This led to contacting forest dieback researchers and subsequently to organizing three international symposia on this topic in 1983, 1987, and 1991.

Each symposium was associated with field trips led by local forest researchers. Observations made during these field trips and interaction with others concerned with the forest dieback/decline syndromes will be illustrated and discussed across a selection of countries visited.

Dieback in Pacific forests

New Zealand

During trips in 1981 and 1983, a number of forest diebacks were noted in New Zealand *Metrosideros* forests. The first

INTERNATIONAL MEETINGS ON FOREST DIEBACK

The first forest dieback symposium took place in 1983 in the frame of the 15th Pacific Science Congress in Dunedin, New Zealand. It was thereafter published in Pacific Science 1983 vol. 37 no. 4 under the theme **Canopy Dieback and Dynamic Processes in Pacific Forests**, with 19 contributions including 7 from Hawai'i, 8 from New Zealand, 3 from Australia and one from Papua New Guinea. The exchange of research findings in the Pacific led to more contacts with forest dieback/decline researchers in Australia, Sri Lanka, Galápagos, Europe, the US, Canada and Japan. A second international dieback symposium was arranged in Berlin at the 14th International Botanical Congress in 1987. The overall congress theme was **Forests of the World**. Our symposium, published in 1988 in *GeoJournal* vol. 17 no.2 had the theme **Stand-level Dieback and Ecosystem Processes: A Global Perspective**. It drew 23 contributions from countries including New Zealand, Australia, the USA, Canada, Sri Lanka, Japan, and Germany. This was followed by a third symposium in 1991 at the 17th Pacific Science Congress in Hilo, HI. This symposium was published in a Springer book 1993 under the title **Forest Decline in the Atlantic and Pacific Regions** with 28 contributions from countries including France, Switzerland, the USA, Germany, Canada, New Zealand, Australia, Papua New Guinea, and Bhutan. (More details in Appendix C.)

was in a northern Rata forest (*Metrosideros robusta*) at Lower Hut, a forest research experiment station near Wellington. The trip resulted from a talk about the 'Ōhi'a forest dieback in Hawai'i given (by Dieter Mueller-Dombois) in a Christmas seminar 1981 at the Victoria University at Wellington. An energetic discussion followed the presentation. One of the scientists, Bob Brockie stood up in the audience and said: "You must have the possum in Hawai'i or some other introduced herbivore. In New Zealand, *Metrosideros* forests are severely attacked and damaged by the introduced Australian Possum [*Trichosurus vulpecola*]." Bob took Dieter Mueller-Dombois the next day to the experiment station and forest at Lower Hut and pointed to a number of defoliated Rata trees that had been observed over many nights from a high platform in a neighboring tree to be completely defoliated by possums. This interaction can be considered biotic dieback (Photo 8.1).

The next encounter with Rata dieback was on the south island in Westland National Park at Franz Joseph Glacier. On the road leading to the glacier there was a sign pointing to the diagonal line on the slope behind saying 1750, implying the glaciers stand at that time (Photo 8.2).

The forest is composed of two dominant canopy trees, and is thus known as the Rata-Kamahi (*Metrosideros umbellata-Weinmannia racemosa*) forest. In this photo the forest above the diagonal line is the older forest in a senescing life stage. The forest below has the same two species as canopy dominants, but it is the younger forest and thus in a vigorous life stage.

Photo 8.1. The Australian possum has become a serious problem in the southern Rata (*Metrosideros umbellata*) forest in New Zealand. This photo shows one of these animals that is being tamed and hand fed by a researcher.

Photo 8.2. The diagonal border seen on the slope indicates the stand of the Franz Joseph glacier in 1750 with Rata-Kamahi senescing forest above and younger forest of the same type below. Westland National Park, January 1982.

Since the possum was considered the sole damaging agent causing dieback in New Zealand's Rata forest, Mueller-Dombois asked at the Westland Park's Headquarters if the diagonal line on the slope was the result of a fence that excluded the possum from the lower healthy forest. The park naturalist Garry McSweeney smiled on hearing this question, went upstairs into the library and pulled out two unpublished reports. One was by an entomologist who claimed that the upper dying forest was hit by an insect pest organism, native to New Zealand, the other report was by Tom Veblen and Glenn Stewart claiming that there was an underlying geological cause of the dieback. That was an introduction to an interesting controversy that was further discussed in our first dieback symposium published in *Pacific Science* in 1983 (referred to above in the blocked statement).

In a subsequent personal discussion with Peter Wardle (author of *Vegetation of New Zealand*), his explanation for the younger Rata-Kamahi forest not being attacked by the Australian possum was that they seem to avoid the younger trees because they can't get a good foothold in the branch system of the younger trees. In the symposium discussion, the question was asked: are the leaves of senescing Rata trees perhaps more palatable than those of the younger, more vigorous trees? A feature common to ʻŌhiʻa in Hawaiʻi was also noted: after a disturbance, in this case an eroded slope segment, southern Rata regenerates in the form of cohorts seen clearly in the center of the slope (Photo 8.3).

A trip to the young volcanic island Rangitoto in Auckland Bay was another eye opener. Here, the third New Zealand *Metrosideros* forest forming tree, pohutukawa (*M. excelsa*) is the monodominant canopy tree of a pioneer forest that resembles pioneer ʻŌhiʻa forests in Hawaiʻi. The Australian possum was not introduced to this island, yet stand-level dieback was noted here as well (Photo 8.4).

Photo 8.3. A juvenile cohort of Rata-Kamahi forest can been seen in the center of this photograph taken in 1982 in Westland National Park, New Zealand. The tree with the red flowers in the lower left hand side of the photo is Rata (*Metrosideros umbellata*).

Photo 8.4. Pohutukawa (*Metrosideros excelsa*) dieback on Rangitoto, a small young volcanic island in Auckland Bay, New Zealand. C.C. Thompson is standing in this photo taken in 1983.

A trip through Cragieburn Forest Park clearly showed a forest dieback in the monospecific mountain beach (*Nothofagus solandri* var. *cliffordioides*) forest. Fortunately, the local research ecologist Udo Benecke was there. He explained that his forest pathologist colleagues believed this dieback to result from a complex interaction of two biotic agents, a fungus and an insect. However, in the 1983 Pacific Science symposium John Wardle and Rob B. Allen explained the mountain beech dieback as resulting from a chain reaction of abiotic causes (Photo 8.5, 8.6, and 8.7).

In New Zealand we encountered the same two different dieback interpretations (biotic vs. abiotic) as initially prevailing in Hawai‘i. The two-cohort feature is shared with the ‘Ōhi‘a replacement dieback in Hawai‘i, and also in the following examples.

Photo 8.5. Mountain beech (*Nothofagus solandri* var. *cliffordiodes*) dieback in 1982 in Cragieburn Forest Park, New Zealand.

Photo 8.6. Regeneration of mountain beech (*Nothofagus solandri var. cliffordioides*) under a forest that experienced canopy dieback in New Zealand in the 1980s; this was a replacement dieback.

Photo 8.7. Mountain beech stand in New Zealand consisting of both senescing old and healthy juvenile cohorts.

Galápagos

On the island of Santa Cruz in the Galápagos islands is a moist mid-elevation forest belt occupied by *Scalesia pedunculata*. It is a fast growing tree in the sunflower family (Asteraceae) that apparently evolved from a weedy colonizer along with sixteen other species, most of which became shrubs. These other woody plants of the genus *Scalesia* occur in different climatic zones throughout the Islands. The tree *Scalesia pedunculata* is of special interest in the Galápagos, because it behaves exactly like our ʻŌhiʻa tree in Hawaiʻi. It also forms monospecific canopies composed of cohort stands. A major difference is its short life cycle of only about 20 years (Itow & Mueller-Dombois 1988) (Photo 8.8, 8.9, and 8.10).

Rain is extremely heavy at times during El Niño events in the Galápagos. This can trigger stand-level dieback when cohort segments of the monospecific canopy are predisposed to die

Photo 8.8. A mature *Scalesia pedunculata* forest at Los Gemeros on Santa Cruz Island, Galápagos, 1981. Photo courtesy of Syuzo Itow.

Photo 8.9. Remnant *Scalesia pedunculata* trees left after this forest on Santa Cruz Island in the Galápagos experienced stand-level dieback during the 1982–1983 El Niño event. In this photo taken in 1987, the understory is dominated by a new cohort of saplings of this same species.

Photo 8.10. Professor Syuzo Itow standing next to a young cohort of *Scalesia pedunculata* regenerating on Santa Cruz Island, Galápagos, in 1987.

from having reached their senescing life stage. Even though the *Scalesia* tree has a much shorter life span and the dieback trigger causes may be different, the response is similar to the canopy dieback process in the Hawaiian rainforest (compare Fig. 6.1 Model A and Fig. 6.2 Model B, p. 142, in Chapter 6 to Fig. 8.1 below).

From top to bottom, the curves in Fig. 8.1 show a 4 year old population, followed by an advanced juvenile cohort, then a mature population, followed by dieback with a remnant senescing group and a new regeneration wave, which is further advanced into a juvenile cohort at the bottom diagram, while the old stand is almost gone; based on data collected by Syuzo Itow in 1988.

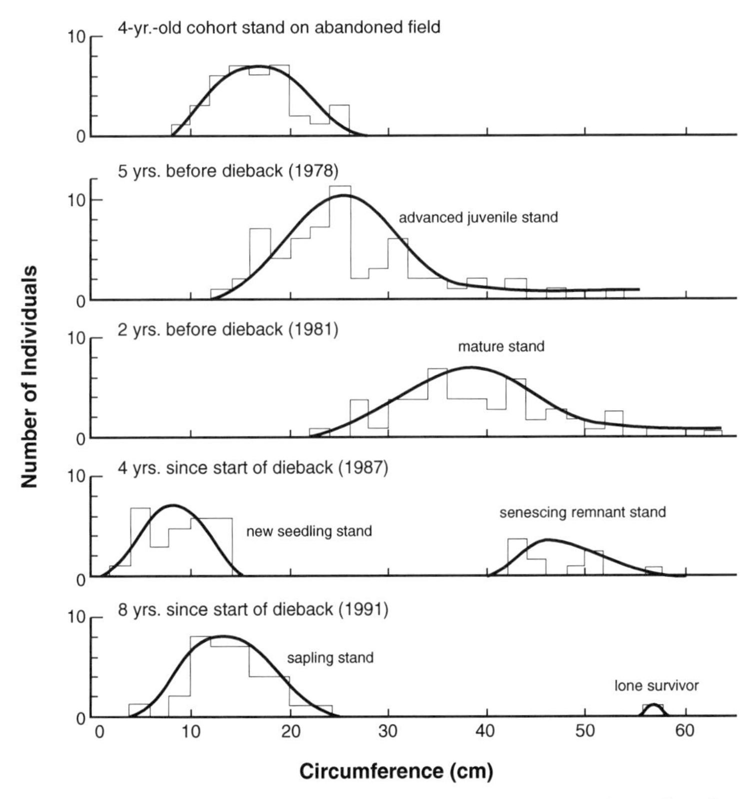

Fig. 8.1. Successive *Scalesia* population curves based on size distribution at Los Gemeros on Isla Santa Cruz.

Japan

The Shimagare phenomenon on Honshu Island in central Japan is also ecologically similar to the other dieback examples that have been described. The name Shimagare refers to the mountain with the "white stripes" of dead or dying fir trees (*Abies* spp.) (Photo 8.11, Fig. 8.2).

One important environmental difference to the tropical examples of Hawai'i and the Galápagos is that the Shimagare phenomenon occurs in a cool temperate environment, similar to that of the mountain beech dieback in New Zealand. The difference to the New Zealand examples is the regularity of structural pattern of the Shimagare phenomenon. The stripes are made up of rows of dead and senescing cohorts of fir trees that often run approximately along contour lines. The intervals between the stripes of dead and dying cohort stands are filled with waves of juvenile firs starting in the wind shelter of the dead and dying

Photo 8.11. View of Shimagare Mountain, Japan, showing the white "stripes" of dead subalpine fir (*Abies* spp.) trees.

parental tree rows in life cycle fashion from seedlings to mature trees. However, the dying fronts do not always follow contour lines; sometimes assuming vertical, wavy patterns of different shapes. (Photo 8.12 and 8.13)

The explanation given for these stripe patterns is that they are initiated by typhoons which cause widespread destruction of the forest across the landscape and that in the following quiet episodes the forest recovers. The most mature front thereafter becomes exposed to the frosty winter storms that trigger the collapse when trees approach their senescing life stage (Photo 8.14 and 8.15).

No biotic agent is involved in the Shimagare fir dieback (Kohyama 1988). The dieback affects two subalpine fir species, *Abies veitchii* and *A. mariesii.* It is of interest here to note that a similar pattern of fir dieback has been observed as "wave regeneration" on White Face Mountain in upper New York State

Photo 8.12. A view within one of the stripes of dead subalpine fir trees on Shimagare Mountain, Japan. Photo by Takashi Kohyama.

Photo 8.13. Inside another Shimagare dieback stand showing the vigorous regeneration of subalpine fir under the dead trees. This stripe of regeneration is approximately 60 m (180 ft) away from the wave front that defines the next dieback stand.

Photo 8.14. Partial displacement dieback on Mt. Shimagare, Japan, where grasses and other introduced species have invaded the understory and restrict natural regeneration of subalpine fir following canopy dieback.

Photo 8.15. View of canopy dieback on Mt. Shimagare, Japan, from a cable car in September 1990.

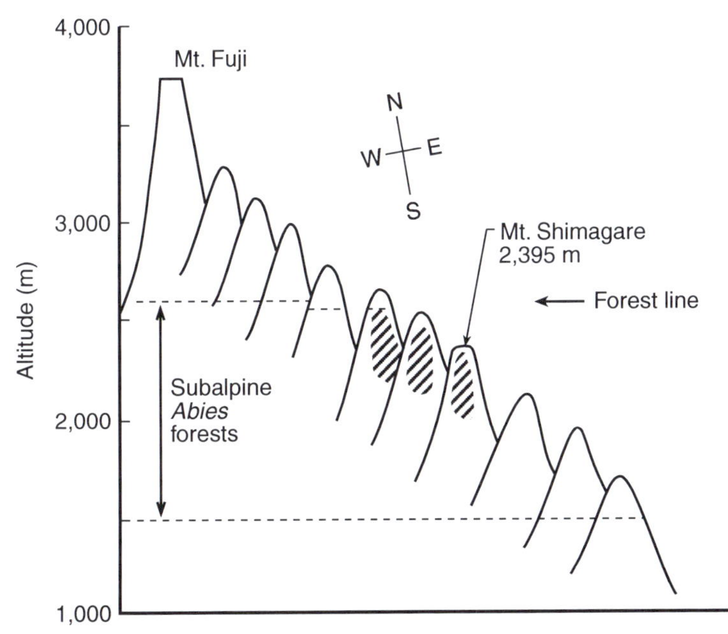

Fig. 8.2. Altitudinal position of subalpine fir forest on Mt. Shimagare relative to Mt. Fuji on Honshu Island, Japan. Courtesy of Takashi Kohyama.

(Spruegel 1977). Therefore, this is a mountain-related as well as an island-related pattern, and low diversity of canopy species and cohort stand structure are underlying factors in all of these stand- or landscape-level breakdowns involving auto-succession, i.e., natural replacement with the same canopy species.

Forest Dieback in the Atlantic Region

Europe

In the 1980s air pollution was considered a threatening cause of forest dieback in Europe; in Germany it was called "Waldsterben." The 1987 IBC (International Botanical Congress) in Berlin had the theme "Forests of the World" in part because of the perceived threat of **Waldsterben.** Acid rain, caused by industrial pollution of sulfur dioxide (SO_2), enhanced levels of toxic nitrogen (NO_X) in the air and ozone (from automobile exhausts) were among the factors blamed for forest dieback noted all over Europe as well as in the eastern United States.

A three week field trip, starting in Switzerland going through Germany's Black Forest, the Czech Republic, the Harz and Egge mountains in northwest Germany, and ending in Berlin, was organized prior to the IBC. It was attended by an international group of forest dieback researchers, including two from Hawai'i (Grant Gerrish and Dieter Mueller-Dombois).

Germany

Here are some photos that allow for an insight on what was going on. Except for the excursion leaders, the participants were mostly from countries outside Germany. Entering the Black Forest, we were surprised about how little damage there was. The media had certainly exaggerated the air pollution damage. We expected to see stands of dead trees such as seen in the Hawaiian rainforest. Instead, the Black Forest spruce stands seemed hardly affected (Photo 8.16, 8.17, and 8.18).

Photo 8.16 (left). Selective tree crown dieback in a European spruce (*Picea abies*) stand, in the Black Forest of Southwest Germany, July 1987.

Photo 8.17. An occasional tree with top dieback can be seen in this photograph taken along a forest trail in same area of the Black Forest, southwest Germany. Surprisingly, no real stand-level dieback is found in this area as is seen in Pacific island forests.

Photo 8.18. A poster describing "Waldsterben" or dying forest at Sirnitz, Black Forest, Germany in August 1985.

THE WALDSTERBEN POSTER

The sign at Sirnitz translates as follows:

This Forest is Dying

This 140 year old spruce forest on the way to Kaelbelescheuer will be dead in a few years. Many of the trees have lost more than 60% of their foliage already now. The remaining needles show a distinct yellowing.

Also young trees are affected.

Air pollution is seen as the main cause, in particular toxic gases such as sulfur dioxide and nitric oxide. Contributing factors are attributable to negative influences of weather, animals and plants.

If we want to save the forest, we need to drastically curtail air pollution caused by industry and automobile traffic. Reduction of toxic gases must be done rapidly and sufficiently.

Please consider your relationship to our common environment. Discuss this problem with the authorities in industry and government.

We all need the forest.

Trees in the forest were marked with different colors as an estimate of their demise in a few years. The estimate was based on current loss of needles.

The cross-section of the spruce trunk on Photo 8.19 also was considered to show a demise of a spruce tree as caused by air pollution, since the annual growth rings show a radical decline around 1980. However, such decline can also be perfectly natural.

The next set of photos show further concerns that revealed themselves during the European forest deline excursion (Waldsterbenexcursion, July 1987).

Photo 8.20 shows the typical spruce forest management in Germany, which consists of raising and harvesting cohort stands. It is reminiscent of the wave regeneration dynamics as found in some Pacific forests under natural conditions, for exam-

Photo 8.19. Cross-section of a spruce trunk showing the size of the tree rings decrease outward, indicating a decline in growth of the tree with age.

Photo 8.20. Cohort forest management in the Egge Mountains in northwest Germany.

ple, the mountain beech forest in New Zealand, the Shimagare phenomenon in Japan, the *Scalesia* forest in the Galápagos, and the ʻŌhiʻa lehua rainforest in Hawaiʻi.

Different hypotheses were proposed by forest dieback researchers to explain the European forest decline. In addition to acid rain, Prof. E.-D. Schulze (1989) suggested that enhanced ammonia (NH_3 = nitrogen in form of gas) precipitates as ammonium salt (NH_4) in forests near agricultural fields thereby over-fertilizing nearby forests. This would disturb the soil nutrient balance, causing decline. On Photo 8.21 he examines small roots of a fallen spruce tree for damage from aluminum toxicity, which could be a factor due to acidification. Grant Gerrish, a member of the Hawaiʻi dieback research team is looking on (Photo 8.21 at right).

Prof. Bernhard Ulrich (1990) was the main proponent of the acid rain hypothesis. On Photo 8.22 he points to a poster

Photo 8.21. The root collar of a windblown tree in the Fichtel Mountains in south central Germany. Prof. E.-D. Schulze (center) points to roots presumably killed as a result of aluminum toxicity, while ʻŌhiʻa dieback researcher Grant Gerrish looks on.

Photo 8.22. Prof. Bernhard Ulrich explaining soil degradation due to acid rain in the Harz Mountains in north central Germany. Dr. E. Matzner holding the poster.

showing that Hydrogen ions have displaced the essential nutrients of Calcium, Magnesium, and Kalium (= Potassium) to such an extent that the soil nutrient budget is severely diminished. He predicted that the surrounding spruce forest will die in less than 10 years. Fortunately these predictions have not come true. Prof. Otto Kandler led the dieback excursion in the fir forest of Bohemia (Photo 8.23). He points out the yellowing needles and considers the forest decline as a complex disease. He was an exception among the European dieback researchers in that he did not consider air pollution an important factor. We started with him as excursion leader in the city of Munich, where he pointed to the healthy fir tree in the densest traffic areas of the city. He asked how can Waldsterben be attributed to poisonous auto emissions, when the trees in the center of the city are growing without any negative effect. He later wrote a paper with the title "The German Forest Decline Situation: A Complex Disease or a Complex of Diseases" (Kandler 1992).

Photo 8.23. Prof. Otto Kandler in July 1987 explaining silver fir (*Abies alba*) dieback in the Bavarian Mountains, Germany, as a complex disease.

On Photo 8.24 Prof. Heinz Zoettl points to a poster showing that Magnesium is in short supply causing spruce tree decline. This, however could have other reasons such as several spruce tree rotations on the same soils without replenishing the lost amount of essential nutrients. R. F. Huettl, who experimentally clarified the loss of Magnesium as a decline factor is looking on. We later collaborated in Hawai'i on the natural 'Ōhi'a dieback, and he realized the similarity of European cohort forest management with the natural cohort dynamics in Hawai'i's 'Ōhi'a rainforest.

However, an important difference is that under natural conditions, without harvesting of the tree biomass organic matter remains at the site as a reserve of nutrients, while under cohort forest management nutrients are extracted with each harvest of the trees.

Photo 8.24. Prof. Heinz Zoettl explains forest dieback due to magnesium deficiency in the Black Forest, Southwest Germany, with Reinhard Huettl looking on. They organized the IUFRO symposium mentioned in Appendix C, p. 257.

Czech Republic

The forest decline excursion continued with a visit into the Giant and Ore Mountains in the Czech Republic and at the border to Poland and the DDR. Here forest decline in the sense of "Waldsterben" was seen as an absolute reality. Photo 8.25 shows the excursion group viewing down a valley that receives a steady stream of sulphur dioxide fumes from the nearby browncoal industry. A vortex of visibly contaminated air is constantly moving into the valley (Photo 8.26). Photo 8.27 shows the spruce tree front of mature trees killed by air pollution. Note that the young growth of spruce saplings in the foreground is still in rather healthy condition.

Near ground the fumes may be less dense, thus the young growth may be more vigorous. Photo 8.28 shows a close-up in the same forest where the dead and dying tree bark is occupied

Photo 8.25. Excursion with Prof. J. Janek into Giant Mountains, a territory severely damaged by dense fumes of sulfur dioxide caused by air pollution.

Photo 8.26. A dense vortex of sulfur dioxide fumes descend into a valley dominated by tall spruce trees. Spruce regeneration in the foreground and taller trees in the background appear to be affected by this pollution and are dying.

Photo 8.27. Planted stand of Norwegian spruce (*Picea abies*) that was killed by concentrated and persistent sulfur dioxide fumes. However, the young trees in the understory appear to be surviving in this photograph taken in 1987.

Photo 8.28. The same dead and dying spruce stand seen in Photo 8.27 with Sirkka Sutinen pointing to lichens on the bark of a dead spruce tree in July 1987.

by lichens. This is rather puzzling, since lichens are known to disappear as the first organisms from the bark of trees in areas of heavy air pollution.

Photo 8.29 shows the excursion members walking through an area of complete die-off in the Giant Mountains. Here there is no doubt about the killing force of air pollution. But it is close to the emission source. Such radical dieback near sulphur dioxide emission sources had been known long before the "Waldsterben" scare in Europe had become a political football. Photo 8.30 shows a less strongly affected location in the same area where young growth was still surviving at that time in July 1987.

We can conclude that forest dieback caused by air pollution was certainly real in areas of intense near-source fumes. But the notion that air pollution emitted through raised chimneys from industrial centers would kill forests in the more remote mountains of central Europe turned out to be wrong. Thorough analyses of forest growth trends in 12 European countries have shown that during the Waldsterben scare the forests outside direct fume areas showed even enhanced growth rates (Spiecker et al. 1996).

The finding of enhanced growth during the decline research episode caused another media bonanza blaming "bad science." This, however, is completely unwarranted. European forest science revealed a great deal of new information about forest growth, for example among many other facts that foliage loss in evergreen forests is not necessarily a predisposition to dieback and that abiotic diebacks are commonly caused by nutritional imbalances.

Photo 8.29. A view of a 130 year old spruce (*Picea abies*) forest killed by sulfur dioxide fumes from pollution caused by the use of brown coal nearby in the Giant Mountains of the Czech Republic. Most of the large and small trees appear to be either dead or dying at this site.

North America

The acid rain hypothesis of forest dieback took hold also in the eastern United States. Dr. R. P. Scheffer, a US plant pathologist and participant in the European forest decline excursion, kindly provided two slides.

The forest dieback of Red spruce and Fraser fir on Balsam Mountain, North Carolina in south Appalachia (Photo 8.31) looks strikingly similar to the air pollution killed spruce forest in the Czech Republic (Photo 8.27 and 8.28). But similar symptoms do not necessarily indicate similar causes. Dr. Scheffer was also not convinced of air pollution as the major cause of the dieback here and on Mt. Mitchell (Photo 8.32). He considered both complete forest die-offs due to a complex of diseases.

Photo 8.30. Another view of the spruce forest affected by pollution in the Giant Mountains of the Czech Republic. However, in this area some of the young spruce appear to be surviving.

Photo 8.31. Dead red spruce and Fraser fir trees as seen in June 1986 on Balsam Mountain, North Carolina which is in southern Appalachia of the United States. Photo by R. P. Scheffer.

Photo 8.32. Another view of die-off of Red spruce and Fraser fir in the Mt. Mitchell area of southern Appalachia in June 1986. Photo by R. P. Scheffer.

Back in Hawai'i

International excursions

European Waldsterben researchers came to Hawai'i for the 1991 symposium in Hilo. Here they were exposed to the different 'Ōhi'a diebacks, Dryland, Wetland, 'Ōhi'a displacement, Bog formation on Maui, and Gap-formation dieback on O'ahu in the northern Ko'olau mountain range at Pūpūkea.

The following six photos depict scenes from international excursions in the Hawaiian Islands that drew forest dieback researchers and ecologists from various countries (Photo 8.33 through 8.38).

Photo 8.33. 'Ōhi'a gap formation dieback in the Pūpūkea section of the northern Ko'olau Mountain Range on the island of O'ahu, Hawai'i. Prof. Heinz Zoettl, pictured here, from Germany was visiting this site in 1990 with his Hawaiian colleagues.

Photo 8.34. Gap-formation dieback on the side of a hill in October 1985 in the Pūpūkea section of the Ko'olau Mountain Range, on the island of O'ahu.

Photo 8.35. Another view of gap-formation dieback in a moist ravine in the Pūpūkea section of the Ko'olau Mountain Range on the island of O'ahu. Grant Gerrish, pictured here in October 1985, studied the Pūpūkea forest in detail for his MSc thesis at the University of Hawai'i.

Photo 8.36. Participants on the IAVS (International Association of Vegetation Science) excursion in 2004 to the Alaka'i bog on the island of Kaua'i. A low-growing shrub variety of 'Ōhi'a (*Metrosideros polymorpha var. pumila*) with red flowers is seen growing in the bog in the right foreground.

Photo 8.37 (left). Along the Pihea trail on the island of Kauaʻi, Prof. Dieter Mueller-Dombois points out to the IAVS excursion participants the ʻŌhiʻa variety (*Metrosideros polymorpha var. dieteri*) that was named in his honor.

Photo 8.38. In 2004 the IAVS excursion also stopped to look at an example of a Kauaʻi dryland ecosystem where the endemic Wiliwili (*Erythrina sandwicensis*) trees seen upslope on the left had been attacked by an introduced gall wasp. In 2008 a parasitic wasp from East Africa was introduced as a biological control agent to help alleviate this problem. It now (2012) appears that many of the Wiliwili trees throughout the Hawaiian Islands have recovered from this problem.

Volcanic fuming (VOG)

A second natural cause of abiotic dieback has manifested itself in form of intensive volcanic fuming. The dieback occurs only near the fuming source. In that case it parallels the anthropogenically caused SO_2 fuming observed in the Giant and Ore Mountains of Europe in the 1980s. Intense volcanic fuming likewise can kill ʻŌhiʻa trees in all size groups as recently discovered at the Chain of Craters Road in Hawaiʻi Volcanoes National Park (Photo 8.39 and 8.40). The acronym VOG stands for volcanically polluted fog.

Photo 8.39. Dense fumes of sulfur dioxide (VOG) are engulfing this juvenile ʻŌhiʻa rainforest near an active volcanic vent along the east rift of Kīlauea volcano in Hawaiʻi Volcanoes National Park. If the SO_2 fuming ceases at the end of the eruption, the trees may recover if they are still alive. Photo by J. Jacobi.

Photo 8.40. Volcanic fume damage seen in January 2012 along the Chain-of-Craters Road in Hawai‘i Volcanoes National Park. Here ‘Ōhi‘a trees have died from intensive and stagnating natural SO_2 fumes, locally called VOG. Small vigorous trees in the crater were killed as well. ‘Ōhi‘a is known to survive volcanic fumes by closing their stomata, but when combined with drought as in this situation, the leaves wilt and start to drop.

Conclusion

Europe’s cohort forest plantation management is not unlike the natural cohort forest mosaics in several ways. Low canopy species diversity is a feature common to both.

Low canopy species diversity leads to more rapid nutrient impoverishment and fragility. It also leads to more limited resource utilization by the narrower spectrum in species life forms and functions.

However, equally important is the resilience of canopy species in terms of their chances for rebirth. Dieback is not to be confused with violent forest mortality, such as caused by a fire, a hurricane or tornado, a volcanic explosion or landslide.

Rebirth of a new cohort forest is also expected to follow non-violent dieback/decline, if the predisposition is due to natural factors and not to a serious disease or site change. Several factors, natural and human-caused, may predispose a forest to dieback. Abiotic dieback is usually caused by nutrient limitations that develop into chronic stresses in the advancing life cycle (premature cohort senescence) of the leading canopy trees to a point of stand-level collapse. The collapse may be triggered by minor climatic disturbances, such as storms that remove a substantial amount of crown foliage from the otherwise evergreen trees, such as the ʻŌhiʻa lehua in Hawaiʻi. The same storm will result in foliage recovery of cohort stands that are in a more vigorous life stage.

The well-known Scottish vegetation ecologist Charles Gimingham (1988) concluded in his synthesis of our second international dieback/decline symposium in Berlin that "forest dieback/decline can not be explained simply as the effect of this or that pollutant. Assessment must be made of the extent to which death from natural causes is to be expected on the basis of natural population dynamics."

A global overview on forest decline and dieback was presented in a 1994 FAO Forestry Paper by two forest pathologists, William Ciesla and Edwin Donaubauer. In addition to the countries cited in this final chapter, they discuss case examples of forest dieback/decline in China, Bhutan, India, Sri Lanka, Bangladesh, Africa (nine cases), Latin America and the Carribean. They found the three stage explanation of **predisposing** factors**, inciting** factors, and **contributing** factors as first suggested by Sinclair (1965), applicable in all cases examined. They also found the cohort senescence theory widely applicable in addition to Manion's (1980) decline theory, which uses a death spiral as a conceptual model. Sinclair and Hudler (1988) likewise gave equal weight to both interpretations.

The ultimate test of which model applies in a particular case should be from what comes after dieback, i.e., the form of recovery.

Global change attributable to human activity provides for new dimensions to the dieback/recovery phenomenon seen worldwide in two ways: (1) Increased opportunities for invasion of human introduced species and human caused changes in landscapes, such as forest to pasture; (2) Change in atmospheric chemistry and global warming and climatic extremes, all of which may affect forest dynamics.

Invasive species have been discussed to some extent in the preceding chapter on fragility versus resilience. However, climate change adds another complication that cannot be clearly predicted in terms of its effect on forest dynamics in the future. We can recognize two components relative to the potential impacts of global climate change:

1. Change in atmospheric chemistry, i.e., carbon dioxide (CO_2) is enhanced, which is a major plant nutrient, and in some areas of industrial agriculture, there is enhanced nitrogen (in form of NH_3) gas that precipitates in forests in form of salt as ammonium nitrogen NH_4, also a major nutrient. Both nutrient enhancements can lead to faster growth rates and thus shortening of growth cycles and also to imbalanced nutrient compositions for forest growth, and thus to factors of predisposition for dieback.
2. Global warming will accelerate plant metabolism and decomposition of organic matter. It thus may enhance forest growth overall, but not necessarily resulting in the same successional patterns as has been experienced

by the Hawaiian ecosystems since ʻŌhiʻa first colonized these islands.

The warming of a few degrees predicted by the Global Climate Models has been experienced before by ʻŌhiʻa lehua in its five million year history of existence in Hawaiʻi. Global warming thus cannot be expected to damage this tree species itself, but may increase the impacts of invasive species on the ʻŌhiʻa forest ecosystem. However, global warming enhances climatic extremes that may trigger forest dieback when trees are predisposed to die. Thus, we consider predisposition, in particular the life-stage related vitality, as the key to understanding stand-level dieback in healthy forest ecosystems.

Ciesla and Donaubauer (1994) conclude their global analysis of forest dieback/decline with a citation, which states: "For evaluating the impact of new anthropogenic stresses such as air pollution, climate change and biotic impoverishment on forests, it is important to understand the natural processes of forest dynamics. Only then will it be possible to untangle the real impact of human influences on forest decline and dieback" (from Mueller-Dombois 1992:36, A natural dieback theory, cohort senescence, as an alternative to the decline disease theory).

Suggested Readings

Allen, C. D. (2009). Climate-induced forest dieback: an escalating global phenomenon? *Unasylva* 60(231/232): 43–49.

Boehmer, H. J. (2011). Vulnerability of tropical montane rain forest ecosystems due to climate change. In *Coping with Global Environmental Change, Disasters and Security – Threats, Challenges, Vulnerabilities and Risks*, ed. by H. G. Brauch, Ú. O. Spring, C. Mesjasz, J. Grin, P. Kameri-Mbote, B. Chourou, P. Dunay, J. Birkmann. Vol. 5. Berlin, Heidelberg, New York: Springer-Verlag, Hexagon Series on Human and Environmental Security and Peace. pp. 789–802.

Mueller-Dombois, D. (1988a). Stand-level dieback and ecosystem processes: A global perspective. *GeoJournal* 17(2): 162–164.

Mueller-Dombois, D. (1988b). Towards a unifying theory for stand-level dieback. *GeoJournal* 17(2): 249–252.

Mueller-Dombois, D. (2006). Long-term rain forest succession and landscape change in Hawaiʻi: The Maui forest trouble revisited. *Journal of Vegetation Science* 17: 685–692.

Huettl, R. F. & Mueller-Dombois, D. (eds.) (1993). *Forest Decline in the Atlantic and Pacific Regions*. Berlin, Heidelberg, New York: Springer-Verlag. 366 p.

A symposium volume with significant contributions on forest dieback and decline in different countries.

Galápagos

Lawesson, J. E. (1988). Stand-level dieback and regeneration of forests in the Galapagos Islands. *Vegetatio* 77: 87–93.

Itow, S. & Mueller-Dombois, D. ((1988). Population structure, stand-level dieback and recovery of *Scalesia pedunculata* forests in the Galapagos Islands. *Ecol. Res.* 3: 333–339.

Japan

Kohyama, T. (1988). Etiology of "Shimagare" dieback and regeneration in subalpine *Abies* forests of Japan. *GeoJournal* 17(2): 201–208.

New Zealand

Ogden, J., Lusk, C. H. & Steel, M. G. (1993). Episodic mortality, forest decline and diversity in a dynamic landscape: Tongariro National Park, New Zealand. In *Forest Decline in the Atlantic and Pacific Region,* ed. by R. F. Huettl and D. Mueller-Dombois. Berlin/Heidelberg: Springer-Verlag. pp. 261–274.

A quote from the introduction reads "Due largely to the work of D. Mueller-Dombois and his associates on the decline of *Metrosideros* in Hawaii and reviews of similar forest dieback phenomena elsewhere in the Pacific region, North America and Europe, we now understand cohort senescence as a plant population phenomenon, rather than as abnormal mortality caused by "disease" or "environmental stress."

The article then demonstrates cohort structure in several New Zealand forests and discusses cohort senescence as an explanatory theory.

Wardle, P. (1991) *Vegetation of New Zealand.* Cambridge, NY: Cambridge University Press. 672 p.

In his book, Vegetation of New Zealand (1991: 560), Peter Wardle made the following statement: "Regeneration of r-selected species should occur within canopy gaps, giving rise to discrete cohorts, each approximating to a bell-shaped (unimodal) diameter-distribution curve. As cohorts age, modal diameters shift to larger diameter classes. This might lead to cohort senescence and thereby to further episodes of synchronized regeneration, even in the absence of renewed disturbance." That statement closely resembles the cohort senescence theory as explained by Mueller-Dombois in the 1983 New Zealand symposium (Pacific Science 37(4): 321). It is also illustrated in Figure 8.1 for the *Scalesia pedunculata* forest in the Galápagos Islands and Fig. 6.1 and 6.2 for *Metrosideros* forest in Hawai'i.

Australia

Lowman, M. D. & Heatwole, H. (1993). Rural dieback in Australia and subsequent landscape amelioration. In *Forest Decline in the Atlantic and Pacific Regions*, ed. by R. F. Huettl & D. Mueller-Dombois. Berlin/Heidelberg: Springer-Verlag. pp. 306–320.

Probably the World's most severe forest dieback and decline due to fragmenting the original eucalypt forest into small, often widely separated remnants to make room for sheep pasture. Fertilization with superphosphate to make pastures profitable for raising sheep caused native insects to overwhelm remaining forest fragments with herbivory. A former forested landscape brought totally out of balance by human activity. A book by the above authors "Dieback: Death of an Australian Landscape" 1986 published by Reed Books Pty Ltd. 150 p. gives the full story with many excellent photos.

New Guinea

Enright, N. J. (1993). Group Death of *Araucaria hunsteinii* K. Schumm (Klinkii pine) in a New Guinea rainforest. In *Forest Decline in the Atlantic and Pacific Regions*, ed. by R. F. Huettl & D. Mueller-Dombois. Berlin/Heidelberg: Springer-Verlag. pp. 321–331.

The study relates to the emergent open monodominant layer of the conifer tree (Klinkii pine) over a closed multispecies brodleaved rainforest canopy. The author explores four hypotheses of dieback occurring in this species: (1) Cohort Senescence, (2) Soil Nutrient Limitation, (3) Changing Water-Relations, (4) Pathogens, Disease, Herbivory. In the conclusion the author considers cohort senescence as predisposition and the host-specific termite *Coptotermis elisae* as the trigger for tree death.

North America

Johnson, D. W., Lindberg, S. E. , Van Miegroet, H., Lovett, G. M., Cole , D. W., Mitchell, M. J. & Binkley, D. (1993). Atmospheric deposition, forest nutrient status, and forest decline: Implications of the integrated forest study. In *Forest Decline in the Atlantic and Pacific Regions*, ed. by R. F. Huettl & D. Mueller-Dombois. Berlin/Heidelberg: Springer-Verlag. pp. 66–81.

Sinclair, W. A. (1965). Comparison of recent declines of white ash, oaks, and sugar maple in northeastern woodlands. *Cornell Plantations* 20: 62–67.

Sinclair, W .A. (1967). Decline of hardwoods: possible causes. Proceedings Intern. *Shade Tree Conference* 42: 17–32.

Sinclair, W. A. and Hudler, G. W. (1988). Tree declines: Four concepts of causality. *Journal of Arboriculture* 14 (2): 29–35.

Sprugel, D. G. (1976). Dynamic structure of wave-regenerated *Abies balsamea* forests in the northeastern United States. *Journal of Ecology* 64: 889–911.

Central Europe

Kandler, O. (1992). The German forest decline situation: A complex disease or a complex of diseases. In *Forest Decline Concepts,* ed. by P. D. Manion & D. Lachance. St. Paul, MN: APN Press. pp. 59–84 .

Sterba, H. (1996). Forest decline and growth trends in central Europe: A review. In *Growth Trends in European Forests*, ed. by H. Spiecker, K. Mielikåinen, M. Koehl & J. P. Skovsgaard. Heidelberg, Berlin, New York: Springer-Verlag, European Forest Institute. pp. 149–165.

From Abstract (p.149): "Forest decline was first recognized in terms of needle loss, and this was defined as indicator for decreasing vitality....only few investigations at that time (in the early 1980s) dealt with stand or plot increments. Later observations especially of permanent plots where a second generation of stands was already being observed presented a tremendous increase in growth."

Spiecker, H., Mielikåinen K., Koehl, M. & Skovsgaard, J. P. (eds) (1996). *Growth Trends in European Forests*. Heidelberg, Berlin, New York: Springer-Verlag, European Forest Institute. 372 p.

This book presents the results of 22 forest growth studies from permanent research plots in 12 European Countries for the period of 1950-1990. The majority of these studies showed increasing growth trends thereby contradicting the widely forecasted trend of forest decline due to air pollution. The positive growth trend in terms of diameter increment and height growth are believed to be attributable to a complex of growth enhancing factors, primarily nitrogen deposition and carbon dioxide enhancement.

Oren, R. & Schulze, E.-D. (1989). 4-H nutritional disharmony and forest decline: A conceptional model. *Oekologia* 77: 425–443.

Nihlgaard, B. (1985). The ammonium hypothesis—an additional explanation to the forest dieback in Europe. *Ambio* 14(1): 1–8.

Schulze, E.-D. (1989). Air pollution and forest decline in a spruce (*Picea abies*) forest. *Science* 244(4906): 776–783.

He considered air pollution responsible for soil nutrient imbalances and adds that "Canopy uptake of atmospheric nitrogen in addition to root uptake stimulated growth and caused a nitrogen to cation imbalance to develop: this imbalance resulted in the decline symptoms."

Schulze, E.-D., Lange, O. L. & Oren, R. (eds) (1989). *Forest Decline and Air Pollution.* Springer Ecological Studies 77. Berlin/Heidelberg: Springer-Verlag. 475 p.

Book Abstract: "During the last decade, forest decline has become increasingly apparent. The decline in forest health was often reported to be associated with air pollution. The present study on Norway spruce stands in the Fichtelgebirge (Bavaria, FRG) analyses various processes interacting within forest ecosystems. It covers transport and deposition of air pollutants, the direct effects of pollutants on above-ground plant parts, the responses of soil to acid rain, the changing nutrient availability as well as the accompanying effects on plant metabolism and growth. The role of fungi, microorganisms and soil animals in the decline of these stands is also assessed. The volume is concluded with a synthesis evaluation of the influence of these different factors, and their interactions on forest decline."

A multi-author volume dealing with forest decline due to air pollution. Since the initially observed damage in the Norway spruce (*Picea abies*) forests of the Fichtel Mountains in south central Germany, four research hypotheses received major attention: (1) natural climatic causes and epidemics, (2) direct effects of air pollutants on above ground plant organs, (3) mineral deficiencies and imbalances as a consequence of acid deposition and soil acidification, and (4) various combinations of these factors.

Note Fig. 6 (p. 13): "Because of intensive forest management the spruce trees are more or less even sized." A widely overlooked factor is the European cohort forest management as a predisposing cause of forest decline.

Ulrich, B. (1990). *Waldsterben*: Forest decline in West Germany. *Envron. Sci. Technol.* 24(4): 436–441.

He held that acid deposition from acid rain affects forest ecosystems in three intensifying phases—Phase I: "the base saturation in soil decreases to zero," Phase II: "acid stress in the subsoil changes the depth gradient of the fine root system toward superficial rooting," and Phase III: "the superficially rooting trees suffer more and more under site-specific stressors." This implies toppling of trees with the next wind-storms (Phase IV).

Zoettl, H. W. & Huettl, R. F. (1991). *Management of Nutrition in Forests under Stress.* Reprinted from *Water, Air, and Soil Pollution* 54, 1990. Dordrecht/Boston/London: Kluwer Academic Publishers. 668 p.

This 1991 book resulted from the IUFRO (International Union of Forest Research Organization) Symposium in Freiburg, Germany, organized by Heinz W. Zoettl & Reinhard F. Huettl in 1989. At this time, the hypotheses on the relation of Air pollution/Waldsterben were considerably toned down. The first paragraph on the book cover (back side) states: "For a long time it has been recognized that forests are under stress in many parts of the word. This is true for natural as well as for managed forest ecosystems." (Further details are given in Appendix C, p. 257–259).

Epilogue

In this book the authors of four generations (Dieter Mueller-Dombois, James D. Jacobi, Hans Juergen Boehmer, Jonathan P. Price) give an overview of five decades of intensive research carried out on the native *Metrosideros polymorpha* rainforests in Hawai'i. Central in this book is the succession dynamics, specifically the processes through which the rainforest assembles on terra nova, remains stable, and rejuvenates itself after perturbations.

The research started in the 70's as a result of the dissatisfaction of the principal author, Professor Dieter Mueller-Dombois, with the general state of knowledge at that time regarding the dieback phenomenon. It was reaching colossal spatial proportions in the rainforest territory of Hawai'i Island. Largely through his and his associate's efforts, the dieback problem became a focal point in forest research in the Hawaiian Islands, and beyond, on a truly global scale. Scientific knowledge exploded exponentially on the dieback problem and with it much new information emerged on the dynamics of forests with canopy tree segments belonging to different generations.

This book far exceeds in completeness anything so far available on the key facets of the 'Ōhi'a lehua rainforest and its multidimensional dynamics. The complete story is told in logical sequence starting with the delightfully crafted cultural Preface of Samuel M. 'Ohukani'ōhi'a Gon III. In the main text, high science is merged with a literal style of writing in colloquial English. The complex thoughts, which matured over the decades, take on an apropos familiarity. The reader is introduced patiently to all key aspects, which make the 'Ōhi'a lehua rainforest a spatially complex, dynamic system. The importance of the canonical approach in research, covering all domains of natural complexity, is shown again and again in examples. Only such an approach could prevent ill-conceived theories to take hold of practice and forestall management interferences, which could undermine system stability or prevent system recovery.

I find the scientific strength of the book in the discussions of rainforest dynamics, particularly in the thoughts on process governance in the context of well-delineated domains of intrinsic regulation and extrinsic forcing. The book masterfully weaves the many details on these into a full-fledged, multifaceted dieback theory.

As could be expected the long decades can take their human toll. The book's contents erect a fitting memorial for those contributors who passed away. The book secures a permanent place for the 'Ōhi'a lehua rainforest project also in the annals of Hawaiian studies.

László Orlóci, FRSC
Emeritus Professor of Statistical Ecology
University of Western Ontario, London, Canada

December 2012

Appendix A

Graduate Students in the Botany Department at the University of Hawai'i with MSc/PhD Theses and Post Doctoral Fellows in Vegetation Ecology Who Contributed to the 'Ōhi'a Lehua Rainforest Story

Carol Newell (MSc 1968) (now Carol Young). A phytosociological study of the major vegetation types in Hawaii Volcanoes National Park, Hawaii.

Mohammad A. Rajput (MSc 1968). Tree stand analysis and soil characteristics of the major vegetation cover types in Hawaii Volcanoes National Park.

Kuswata Kartawinata (PhD 1971). Phytosociology and ecology of the natural dry grass communities on Oahu, Hawaii. [Kuswata did the Walter-type climate diagrams shown on Fig. 3.1 and 3.2 in this book.]

Garrett A. Smathers (PhD 1972). Invasion, early succession and recovery of vegetation on the 1959 Kīlauea volcanic surfaces.

Jean Maka (MSc. 1973). A mathematical approach to defining recurring species groups in a montane rain forest on Mauna Loa, Hawai'i.

Ranjit Cooray (MSc. 1974). Stand structure of a montane rainforest on Mauna Loa, Hawai'i.

Richard Becker (PhD 1976). The phytosociological position of tree ferns (*Cibotium* spp.) in the montane rain forests of the island of Hawai'i.

Grant Gerrish (MSc. 1978). The relationship of native and exotic plant species in two rain forest communities in the Koʻolau Mountains, Oʻahu, Hawaiʻi.

Nengah Wirawan (PhD 1978). Vegetation and soil water regimes in a tropical rain forest valley on Oahu, Hawaiian Islands.

Philip J. Burton (MSc. 1980). Light regimes and *Metrosideros* regeneration in a Hawaiian montane rain forest.

Robert A. Wright (MSc. 1985). Shrub population structure and dynamics in a primary successional community, Devastation Area, Hawaiʻi.

Nadarajah Balakrishnan (PhD 1985). Vegetation patterns and nutrient regimes in a tropical montane rain forest ecosystem, Hawaiʻi.

Joan E. Canfield (PhD 1986). The role of edaphic factors and plant water relations on plant distribution in the bog/wet forest complex of ʻAlakaʻi Swamp, Kauaʻi, Hawaiʻi.

R. Lani Stemmermann (PhD 1986). Ecological studies of ʻŌhiʻa varieties (*Metrosideros polymorpha*, Myrtaceae), the dominants in successional communities of Hawaiian rain forests.

R. Alan Holt (MSc. 1988). The Maui forest trouble: reassessment of an historic forest dieback.

Grant Gerrish (PhD 1988). The changing carbon balance in aging *Metrosideros polymorpha* trees.

James D. Jacobi (PhD. 1990). Distribution maps, ecological relationships, and status of native plant communities on the island of Hawaiʻi.

Kanehiro Kitayama (PhD 1991). Comparative vegetation analysis on the wet slopes of two tropical mountains, Mt. Haleakala, Maui, and Mt. Kinabalu, Borneo.

Wayne Takeuchi (PhD 1991). The *Metrosideros polymorpha* forest of ʻAlakaʻi swamp: population structures and dynamic trends.

Dennis Grossman (PhD 1992). Early recovery of lowland rainforest following clear-cutting at Kalapana on the island of Hawaiʻi.

Patricia C. Welton (MSc 1992). Community organization and population structures of a lowland mesic forest, NW Oʻahu.

K. Mallikarjuna Aradhya (PhD 1992). Genecology of *Metrosideros* in an altitudinal and successional perspective.

Donald D. Drake (PhD 1993). Population ecology of *Metrosideros polymorpha* and some associated plants of Hawaiian volcanoes.

Cynthia C. Lipp (PhD 1994). Ecophysiological limits and community level constraints to *Myrica faya* invasion in Hawai'i Volcanoes National Park.

Yoshiko Akashi (PhD 1994). The landscape ecology of *Metrosideros* dieback in Hawai'i.

Vanessa Minden (MSc. 2005). The effects of the invasion of *Hedychium gardnerianum* Ker.-Gaw (Zingiberaceae) on the regeneration dynamics of a montane rainforest on the island of Hawai'i.

Post-Doctoral Fellows in Vegetation Ecology

Guenter Spatz

Spatz, G. (1973). Habil. Dissertation: Ein Nutzungsvorschlag fuer die Keauhou Ranch auf Hawaii aus landwirtschaftlicher und oekologischer Sicht [A Plan for Improving the Agricultural Use of Keauhou Ranchland from an Ecological Perspective] (with vegetation and soil maps, written in German). Technical University of Munich, Freising-Weihenstephan, Department of Landscape Ecology.

Spatz, G. & Mueller-Dombois, D. (1973). The influence of feral goats on koa tree reproduction in Hawaii Volcanoes National Park. *Ecology* 54(4): 870–876.

Spatz, G. & Mueller-Dombois, D. (1975). Succession patterns after pig digging in grassland communities on Mauna Loa, Hawaii. *Phytocoenologia* 3(2/3): 346–373.

Mueller-Dombois, D. & Spatz, G. (1975). Application of the relevé method to insular tropical vegetation for an environmental impact study. *Phytocoenologia* 2(3/4): 417–429.

Grant Gerrish

Gerrish, G. (1988). An explanation of natural forest dieback based on the "pipe model" analogy. *GeoJournal* 17(2): 295–299.

Gerrish, G., Mueller-Dombois, D. & Bridges, K. W. (1988). Nutrient limitations in *Metrosideros* dieback forests in Hawaii. *Ecology* 69: 723–727.

Gerrish, G. & Mueller-Dombois, D. (1999). Measuring stem growth rates and cohort analysis of a tropical evergreen tree. *Pacific Science* 53(4): 418-429.

Kanehiro Kitayama

Kitayama, K. & Mueller-Dombois, D. (1995). Biological invasion on an oceanic island: Do alien species have wider ecological amplitudes than native species? *Journal of Vegetation Science* 6: 667–674.

Kitayama, K. & Mueller-Dombois, D. (1995). Vegetation changes along gradients of long-term soil development in the Hawaiian montane rainforest zone. *Vegetatio* 120: 1–20.

Kitayama, K., Schuur, E. A. G., Drake, D. R. & Mueller-Dombois, D. (1997). Fate of a wet montane forest during soil ageing in Hawaii. *Journal of Ecology* 85: 669–679.

Kitayama, K., Pattison, R., Cordell, S., Webb, D. & Mueller-Dombois, D. (1997). Ecological and genetic implications of foliar polymorphism in *Metrosideros polymorpha* Gaud. (Myrtaceae) in a habitat matrix on Mauna Loa, Hawai'i. *Annals of Botany* 80: 491–497.

Reinhard F. Huettl

Huettl, R. F. (1991). Post-doc. Habil. Dissertation: Die Nährelementversorgung geschaedigter Waelder in Europa und Nordamerika [Nutritional Needs in Forests affected by "New-Type Forest Damage" in Europe and North America]. University of Freiburg, Germany, Freiburger Bodenkundliche Abhandlungen 28. 440 p.

Reinhard Huettl was Visiting Assistant Professor in Geobotany, UHM Department of Botany, during 1991–1992. He gave lectures in two courses: Vegetation Ecology, and Plant Ecology & Environmental Measurements.

On a prior visit, Reinhard Huettl observed 'Ōhi'a dieback on Kaua'i Island. Thereafter, he initiated the research relationship to the European forest decline.

Huettl, R. F. & Mueller-Dombois, D. (eds.) (1993). *Forest Decline in the Atlantic and Pacific Regions*. Berlin, Heidelberg, New York: Springer-Verlag. 366 p.

Hans Juergen Boehmer

Boehmer, H. J. (2005). Post-doc. Habil. Dissertation: Dynamik und Invasibilitaet des montanen Regenwaldes auf der Insel Hawaii [Dynamics and Invasibility of Hawaii's Montane Rainforest]. Technical University of Munich, Germany, Department of Ecology and Ecosystem Managment. 232 p. Includes six appendices.

Boehmer, H. J. & Niemand, C. (2009). Die neue Dynamik pazifischer Wälder. Wie Klimaextreme und biologische Invasionen Inselökosysteme verändern [The new dynamics of Pacific forests. How climatic anomalies and biological invasions change island ecosystems]. *Geographische Rundschau* 61: 32–37.

Boehmer, H. J. (2011). Vulnerability of tropical montane rain forest ecosystems due to climate change. In *Coping with Global Environmental Change, Disasters and Security—Threats, Challenges, Vulnerabilities and Risks* ed. by H. G. Brauch, Ú. Oswald Spring, C. Mesjasz, J. Grin, P. Kameri-Mbote, B. Chourou, P. Dunay & J. Birkmann. Hexagon Series on Human and Environmental Security and Peace, vol. 5. Berlin, Heidelberg, New York: Springer-Verlag. pp. 789–802.

Boehmer, H. J. (2011): Störungsregime, Kohortendynamik und Invasibilität—zur Komplexität der Vegetationsdynamik im Regenwald Hawaiis [Disturbance regimes, cohort dynamics, and invasibility—on the complexity of vegetation dynamics in Hawaii's rainforests]. *Laufener Spezialbeiträge, Landschaftsökologie, Grundlagen, Methoden, Anwendungen* 2011: 111–117.

Boehmer, H. J., Wagner, H.H., Jacobi, J.D., Gerrish, G.C. & Mueller-Dombois, D. (In press, 2013). Rebuilding after collapse: Evidence for long-term cohort dynamics in the native Hawaiian rainforest. *Journal of Vegetation Science.* DOI: 10.1111/jvs.12000.

Appendix B

Checklist of Plants Commonly Found in the ʻŌhiʻa Rainforest

I = **Indigenous:** species that arrived naturally without human help

E = **Endemic:** species that are found only in that geographic area and nowhere else in the world

X = **Introduced:** non-native species; also referred to as naturalized, exotic, foreign, or alien species

Ferns

E *Adenophorus hymenophylloides* (Grammitidaceae) Pai, Palai huna

E *Adenophorus tamariscinus* var. *tamariscinus* (Grammitidaceae) Wahine noho mauna

I *Asplenium aethiopicum* (Aspleniaceae) ʻIwaʻiwa a Kāne

E *Asplenium contiguum* var. *contiguum* (Aspleniaceae)

I *Asplenium lobulatum* (Aspleniaceae) Piʻipiʻi lau manamana, ʻAnaliʻi

E? *Asplenium macraei* (Aspleniaceae) ʻIwaʻiwa lau liʻi

E *Asplenium trichomanes* subsp. *densum* (Aspleniaceae) ʻOāliʻi

E *Athyrium microphyllum* (Athyriaceae) ʻĀkōlea

E *Cibotium glaucum* (Dicksoniaceae) Hāpuʻu, Hāpuʻu pulu

E *Cibotium menziesii* (Dicksoniaceae) Hāpuʻu ʻiʻi, ʻIʻi, ʻIʻiʻi

I *Dicranopteris linearis f. linearis* (Gleicheniaceae) Uluhe, Unuhe

E *Diplazium sandwichianum* (Athyriaceae) Hōʻiʻo, Pohole (Maui)

E *Diplopterygium pinnatum* (Gleicheniaceae) Uluhe lau nui

I *Dryopteris wallichiana* (Dryopteridaceae) ʻIʻo nui, Laukahi

I *Grammitis hookeri* (Grammitidaceae) Mākuʻe lau liʻi

E *Huperzia erosa* (Lycopodiaceae)

I *Huperzia serrata* (Lycopodiaceae)

I *Lycopodiella cernua* (Lycopodiaceae) Wāwaeʻiole, Hulu ʻiole, Huluhulu a ʻiole

I *Lycopodium venustulum* var. *venustulum* (Lycopodiaceae)

I *Microlepia strigosa* var. *strigosa* (Dennstaedtiaceae) Palapalai, Palai

X *Nephrolepis brownii* (Nephrolepidaceae) swordfern

I *Nephrolepis cordifolia* (Nephrolepidaceae)

I *Nephrolepis exaltata* (Nephrolepidaceae) ʻŌkupukupu, Niʻaniʻau, Pāmoho, Kupukupu, Palapalai

X *Nephrolepis multiflora* (Nephrolepidaceae, now in Lomariopsidaceae) False ʻŌkupukupu

E *Pseudophegopteris keraudreniana* (Thelypteridaceae) Waimakanui, ʻĀkōlea, Alaʻalai

I *Pteris cretica* (Pteridaceae) ʻŌali, cretan brake

I *Pteris excelsa* (Pteridaceae) Waimakanui, ʻIwa

E *Pteris irregularis* (Pteridaceae) Mānā, ʻĀhewa (Oʻahu), ʻIwa puakea (Maui)

E *Sadleria cyatheoides* (Blechnaceae) ʻAmaʻu, Maʻu, Maʻumaʻu, Puaʻa ʻehuʻehu, ʻAmaʻumaʻu

E *Sadleria pallida* (Blechnaceae) ʻAmaʻu ʻIʻi, ʻIʻi, ʻIʻiʻi, ʻAmaʻu, Maʻu, Maʻumaʻu, Puaʻa ʻehuʻehu, ʻAmaʻumaʻu

E *Sadleria souleyetiana* (Blechnaceae) ʻAmaʻu, Maʻu, Maʻumaʻu, Puaʻa ʻEhuʻehu, ʻAmaʻumaʻu

I *Sphenomeris chinensis* (Lindsaeaceae) Palaʻā, Palapalaʻā, Palaʻe, Pʻāʻū o palaʻe

E *Sticherus owhyhensis* (Gleicheniaceae) Uluhe, Unuhe

Grasses and Sedges

X *Andropogon virginicus* var. *virginicus* (Poaceae) broomsedge, yellow bluestem

E *Carex alligata* (Cyperaceae)

X *Holcus lanatus* (Poaceae) common velvet grass, Yorkshire fog

E *Isachne distichophylla* (Poaceae) ʻOhe

E *Joinvillea ascendens* subsp. *ascendens* (Joinvilleaceae) ʻOhe

I *Machaerina angustifolia* (Cyperaceae) ʻUki

E *Machaerina mariscoides* subsp. *meyenii* (Cyperaceae) ʻAhaniu, ʻUki

E *Trisetum glomeratum* (Poaceae) Pili uka, Heʻupueo (Hawaiʻi), mountain Pili

I *Uncinia uncinata* (Cyperaceae)

Forbs (herbaceous species)

E *Anoectochilus sandvicensis* (Orchidaceae) jewel orchid

X *Arundina graminifolia* (Orchidaceae) bamboo orchid

E *Astelia menziesiana* (Liliaceae) Kaluaha, Puaʻakuhinia, Paʻiniu

X *Hedychium gardnerianum* (Zingiberaceae) Kāhili ginger, Kāhili, ʻAwapuhi kāhili

E *Liparis hawaiensis* (Orchidaceae) ʻAwapuhiakanaloa

I *Nertera granadensis* (Rubiaceae) Mākole

E *Peperomia cookiana* (Piperaceae) ʻAlaʻala wai nui

E *Peperomia hypoleuca* (Piperaceae) ʻAlaʻala wai nui

E *Peperomia macraeana* (Piperaceae) ʻAlaʻala wai nui

I *Peperomia tetraphylla* (Piperaceae) ʻAlaʻala wai nui

Shrubs

E *Broussaisia arguta* (Hydrangeaceae) Kanawao, Pūʻahanui

E *Clermontia hawaiiensis* (Campanulaceae) ʻŌhā wai nui, ʻŌhā kēpau, ʻŌhāhā wai nui, ʻŌhā wai, ʻŌhā, Hāhā

E *Clermontia montis-loa* (Campanulaceae) ʻŌhā wai, ʻŌhā, Hāhā

E *Clermontia parviflora* (Campanulaceae) ʻŌhā wai, ʻŌhā, Hāhā

E *Cyanea floribunda* (Campanulaceae) Hāhā

E *Cyanea longiflora* (Campanulaceae) Hāhā

E *Cyanea pilosa* subsp. *pilosa* (Campanulaceae) Hāhā

E *Cyanea tritomantha* (Campanulaceae) ʻAkū, Hāhā

E *Cyrtandra lysiosepala* (Gesneriaceae) Haʻiwale, Kanawao keʻokeʻo

E *Cyrtandra paludosa* var. *paludosa* (Gesneriaceae) Moa, Hahala, Haʻiwale, Kanawao keʻokeʻo

E *Cyrtandra platyphylla* (Gesneriaceae) ʻIlihia, Haʻiwale, Kanawao keʻokeʻo

E *Dubautia scabra* subsp. *scabra* (Asteraceae) Kūpaoa, Naʻenaʻe

I *Dodonaea viscosa* (Sapindaceae) ʻAʻaliʻi

E *Kadua affinis* (Rubiaceae) Manono

E *Kadua centranthoides* (Rubiaceae)

I *Leptecophylla tameiameiae* (formerly *Styphelia tameiameiae*) (Epacridaceae) Pūkiawe

X *Miconia calvescens* (Melastomataceae) Miconia

E *Pipturus albidus* (Urticaceae) Mamake

E *Platydesma spathulata* (Rutaceae) Pilo kea

X *Pluchea carolinensis* (Asteraceae) sourbush, marsh fleabane

E *Rubus hawaiensis* (Rosaceae) ʻĀkala, ʻĀkalakala, Kala

X *Tibouchina urvilleana* var. *urvilleana* (Melastomataceae) lasiandra, princess flower

E *Touchardia latifolia* (Urticaceae) Olonā

E *Trematolobelia wimmeri* (Campanulaceae)

E *Vaccinium calycinum* (Ericaceae) ʻŌhelo, ʻŌhelo kau lāʻau

E *Vaccinium reticulatum* (Ericaceae) ʻŌhelo, ʻŌhelo ʻai

Vines

I *Alyxia stellata* (Apocynaceae) Maile

I *Freycinetia arborea* (Pandanaceae) ʻIeʻie, ʻie

E *Stenogyne calaminthoides* (Lamiaceae)

Trees

E *Acacia koa* (Fabaceae) Koa

E *Antidesma platyphyllum* var. *platyphyllum* (Euphorbiaceae) Hame, Haʻā, Haʻāmaile, Hamehame, Mehame, Mehamehame

X *Casuarina equisetifolia* (Casuarinaceae) common ironwood, Paina

E *Cheirodendron trigynum* subsp. *trigynum* (Araliaceae) ʻŌlapa, Lapalapa

E *Coprosma ochracea* (Rubiaceae) Pilo, Hupilo

E *Coprosma rhynchocarpa* (Rubiaceae) Pilo, Hupilo

E *Erythrina sandwicensis* (Fabaceae) Wiliwili

X *Falcataria moluccana* (Fabaceae) Albizia

I *Ilex anomala* (Aquifoliaceae) Kāwaʻu, ʻAiea (Kauaʻi)

E *Kadua axillaris* (Rubiaceae) Manono

E *Melicope clusiifolia* (formerly *Pelea clusiifolia*) (Rutaceae) Kūkaemoa (Kauaʻi), Kolokolo mokihana, Alani, Alani kuahiwi

E *Melicope radiata* (Rutaceae) Alani, Alani kuahiwi

E *Melicope volcanica* (Rutaceae) Alani, Alani kuahiwi

E *Metrosideros polymorpha* var. *dieteri* (Myrtaceae) ʻŌhiʻa, ʻŌhiʻa lehua, Lehua

E *Metrosideros polymorpha* var. *glaberrima* (Myrtaceae) ʻŌhiʻa, ʻŌhiʻa lehua, Lehua

E *Metrosideros polymorpha* var. *incana* (Myrtaceae) ʻŌhiʻa, ʻŌhiʻa lehua, Lehua

E *Metrosideros polymorpha* var. *macrophylla* (Myrtaceae) ʻŌhiʻa, ʻŌhiʻa lehua, Lehua

E *Metrosideros polymorpha* var. *polymorpha* (Myrtaceae) ʻŌhiʻa, ʻŌhiʻa lehua, Lehua

E *Metrosideros polymorpha* var. *pumila* (Myrtaceae) ʻŌhiʻa, ʻŌhiʻa lehua, Lehua

X *Morella faya* (formerly *Myrica faya*) (Myricaceae) firetree

I *Myoporum sandwicense* (Myoporaceae) Naio, Naeo, Naieo, bastard sandalwood

E *Myrsine lessertiana* (Myrsinaceae) Kōlea lau nui, Kōlea

E *Myrsine sandwicensis* (Myrsinaceae) Kōlea lau liʻi, Kōlea

E *Perrottetia sandwicensis* (Celastraceae) Olomea, Puaʻa olomea, Waimea (Maui)

E *Pritchardia beccariana* (Arecaceae) Loulu

X *Psidium cattleianum* (Myrtaceae) strawberry guava, Waiawī ʻulaʻula

E *Psychotria hawaiiensis* var. *hawaiiensis* (Rubiaceae) Kōpiko ʻula, ʻŌpiko

I *Sapindus saponaria* (Sapindaceae) Mānele

E *Sophora chrysophylla* (Fabaceae) Māmane, Mamani

X *Syzygium jambos* (Myrtaceae) rose apple, ʻŌhiʻa loke

E *Urera glabra* (Urticaceae) Ōpuhe, Hōpue, Hona

For pictures of Hawaiian native and non-native plants, visit:

Hawaii Ecosystems at Risk's Plants of Hawaii (now maintained by Forest and Kim Starr at their website): http://www.starrenvironmental.com/images/?o=plants

Smithsonian Institution's Flora of the Hawaiian Islands Checklist: Image Gallery: http://www.botany.si.edu/pacificislandbiodiversity/hawaiianflora/imagegallery.cfm

Appendix C

Research History of the ʻŌhiʻa Rainforest Dieback/Decline: An Annotated Bibliography

Dieter Mueller-Dombois

What is meant by forest dieback/decline?

This term is here defined as "the loss of crown foliage of individual trees without any significant replacement and more broadly as canopy dieback on a stand-level basis." Dieback and decline can be a rapid or a long drawn-out process, but it ends with death of the affected trees. The same sort of dieback/decline can be observed in stands of shrubs. The terms dieback and decline have been used for the same phenomenon.

The background

In 1964, a mountain transect analysis from the summit of Mauna Kea along the windward east slope to the sugar cane fields near Hilo revealed ʻŌhiʻa lehua (*Metrosideros polymorpha*) stand-level dieback in the poorly drained transect segment #9 from 4500–2900 feet elevation (1,370 m down to 880 m) near Wailuku Stream. "Here, the rainfall is very high (5,000 mm/year), fog is almost constantly present, and soil drainage is impeded. It is the area on the Mauna Kea slope where numerous stream beds begin to form, but where they are not yet deeply cut into the substrate." (Mueller-Dombois & Krajina 1968: 517).

Mueller-Dombois, D. & Krajina, V. J. (1968). Comparison of east-flank vegetations on Mauna Loa and Mauna Kea, Hawaii. In *Proc. Symp. Recent Advances in Trop. Ecol.*, ed. by R. Misra and B. Gopal. Vol. II. Varanasi: International Society for Tropical Ecology. pp. 508–520.

This was the first published account on the native 'Ōhi'a (*Metrosideros polymorpha*) forest dieback on the island of Hawai'i that started a wave of research, beginning in the early 1970s.

(See Mauna Kea transect profile, Fig.3.4, and 'Ōhi'a dieback Photo 3.23 in Chapter 3.)

An early and enduring controversy

As the newly appointed Scientific Coordinator & Executive Director of the NSF funded IPB/Hawai'i (the International Biological Program in Hawai'i), I organized a field excursion in August 1970 into that above mentioned dieback territory. The group included about 25 researchers who were interested in participating in the just funded IBP/Hawai'i. Getting into the dieback area resulted in a lively discussion of possible causes. An epidemic disease was immediately predicted and staunchly argued for by Prof. Ivan W. Buddenhagen, Head of the Plant Pathology Department, University of Hawai'i at Mānoa (UHM). Thinking otherwise, I proposed a natural cause as an alternative to the biotic disease idea. My first impression gained during the Mauna Kea transect analysis led me to argue for a natural cause as stated in form of a quote in the above cited paper. At this field gathering, I pointed to trees standing in water, and considered poor soil drainage during a transition phase in geomorphological development on this young volcanic mountain as a possible cause of the dieback.

Subsequently, these preliminary assessments led to major research projects. Following the fieldtrip, Prof. Buddenhagen contacted the USDA Institute of Pacific Islands Forestry and persuaded its Director, Robert E. Nelson to request research funding for this threatening "epidemic disease." Soon thereafter, the 'Ōhi'a Dieback/Decline research started with federal funding given to government forest researchers with contracts to the Beaumont Agricultural Research Center at the University of Hawai'i at Hilo (UHH), the Plant Pathology Department at the UHM, and to the Bishop Museum Entomology Departments. Several entomologists were funded to do insect pest research.

Two major programs running parallel in the 1970s

Now, two federally funded research programs were conducted side-by-side, the IBP/Hawai'i (called Island Ecosystems Stability and Evolution Subprogram) and the USDA Forest Service program focused on the 'Ōhi'a forest disease. The Co-Director of the IBP/Hawai'i, Prof. Andrew Berger, insisted that IBP concentrate on the intact natural ecosystems and leave the problem of forest decline to the US Forest Service.

Disease research on the ʻŌhiʻa forest dieback/decline

The following citations refer to authors who promoted the epidemic disease hypothesis:

Burgan, R. E. & Nelson, R. E. (1972). *Decline of* ʻŌhiʻa Lehua *Forests in Hawaii.* Berkeley, CA: USDA Forest Service Gen. Tech. Rep. PSW-3, Pac. SW Forest and Range Expt. Sta. 4 p.

Burgan and Nelson refer to the historic and first serious disease study of the ʻŌhiʻa forest decline on Maui Island by Harold Lyon:

Lyon, H. L. (1909). The forest disease on Maui. *Hawaiian Planter's Record* 1: 151–159.

However, ten years later, Harold Lyon corrected his first prediction that a killer disease was responsible for the Maui ʻŌhiʻa forest dieback. Instead he emphasized an abiotic cause for the Maui ʻŌhiʻa forest dieback:

Lyon, H. L. (1919). Some observations on the forest problems of Hawaiʻi. *Hawaiian Planter's Record* 21: 289–300.

In this follow-up paper, Lyon suggested that soil toxicity was causing the Maui forest decline/dieback rather than a biotic disease as earlier suspected.

Yet, a new disease was strongly suspected in the new Hawaiʻi Island dieback/decline as proclaimed by the following papers:

Bega, R. V. (1974). *Phytophthora cinnamomi*: its distribution and possible role in ʻŌhiʻa Decline on the Island of Hawaii. *Plant Disease Reporter* 58:1089–1073.

"*Phytophthora cinnamomi* was isolated from soil and rootlet samples. About 175,000 acres (70,000 ha) of the ʻŌhiʻa lehua (*Metrosideros collina ssp. polymorpha*) are in various stages of decline. All evidence suggests that the disease is spreading rapidly."

Petteys, E. Q. P., Burgan, R. E. & Nelson, R. E. (1975). *ʻŌhiʻa forest decline: Its spread and severity in Hawaii.* Albany, CA: USDA Forest Service, PSW-105. 11 p.

This paper demonstrates the rapid decline of native ʻŌhiʻa forest on the east slopes of Mauna Kea and Mauna Loa from aerial photo sets taken in 1954, 1965, and 1972. Progression is shown on maps in an 80,000 ha study area above the Hamakua Coast upslope from Laupahoehoe (about 15 km north of Hilo). Here the dieback begins between 3000–5000 feet elevation (Fig.1, p. 2;

Fig. 2, p. 8 in that paper), from North to South into Hawai'i Volcanoes National Park showing severe decline from initially120 ha in1954 to16,000 ha in1965 and then to 34,500 ha in 1972. "This study suggests that if decline continues at the present rate, remaining 'Ōhi'a forest in the study area will be virtually eliminated in 15–25 years." (Fortunately, this did not become true.)

In an April 1975 letter of Robert E. Nelson (Director of Pacific Islands Forestry Institute, Honolulu) to Dr. Wen-Hsiung Ko (Plant Pathologist at Beaumont Agricultural Experimental Station, University of Hawai'i at Hilo) Nelson stated: "Our working hypothesis on the cause of the decline is that a complex of pathogenic fungi and native forest insects are the principal agents involved in the epidemic decline."

At about the same time Robert E. Nelson promoted this working hypothesis, almost as a prediction in a new 5-year research plan on 'Ōhi'a Forest Decline (April 1975). The plan had three major objectives:

A. Determine extent and severity of forest decline

B. Determine causes of 'Ōhi'a forest decline

C. Develop and test methodology needed to prevent or control 'Ōhi'a forest decline

As tests he suggested: (1) Replanting with exotic and native trees including 'Ōhi'a, (2) Fertilizer tests to control decline, (3) Fungicide tests (4) Insecticide tests, and (5) Biological control tests.

Trial planting in decline and non-decline areas was considered a useful option at this early stage before the decline causes were clarified. Testing with fertilizer, fungicide, insecticide, and biological control was not to be started prior to clarifying the causes.

Intensive and thorough disease research was underway, but it soon turned out to be skeptical of the prediction that a biotic disease epidemic was ravaging through the native 'Ōhi'a rainforest. This became clear through the following papers:

Kliejunas, J. T. & Ko, W. H. (1974). Deficiency of inorganic nutrients as a contributing factor to 'Ōhi'a decline. *Phytopathology* 64: 891–896.

This study was done on the 1855 Mauna Loa lava flow at 4,000 ft elevation where the trees were still only about 100 years old. "Results indicated that 'Ōhi'a trees are declining because of nutrient deficiency."

This study can be considered a special case in the dieback syndrome as it was done on a young, only 120 year old pāhoehoe substrate. Here only small groups and some individuals were in a state of dieback, the majority of the juvenile 'Ōhi'a trees had still well foliated crowns.

Kliejunas, J. T. & Ko, W. H. (1975). The occurence of *Pythium vexans* in Hawaii. *Plant Disease Reporter* 59:392–395.

Results: *P. vexans* is a root rot fungus that caused root rot of 'Ōhi'a in greenhouse and field tests. But there was no correlation found between 'Ōhi'a tree decline and *P. vexans*. Where *P. vexans* was present in decline areas, *Phytophthora cinnamomi* was also present. In greenhouse tests *P. cinnamomi* was the more virulent root pathogen. (Dr. Wen-Hsiung Ko made special efforts to fall in line with the officially declared disease hypothesis, but could not support it as seen in the 1978 publication with his PhD graduate student S. C. Hwang, see below).

Dr. Wen-Hsiung Ko had stepped out of line with his "premature" fungicide and fertilizer tests, and was subsequently dropped from the forestry team in terms of funding. He came to me asking for support of his graduate student S. C. Huang, who had embarked on a study of the soil biology of *Phytophthora cinnamomi* for his dissertation research. He found that this fungus in Hawai'i does not behave as an obligatory pathogen but rather as a saprophyte. It's relationship is with moist soil moisture regimes, not wet or dry soils. It also is likely an indigenous member among Hawai'i's soil organisms. These findings were published in a joint paper cited here:

Hwang, S. C. & Ko, W. H. (1978). Quantitative studies of Phytophthora cinnamomi in decline and healthy 'Ōhi'a forests. *Trans. British Mycol. Soc.* 70(2): 312–315.

Because of the strong belief among research foresters and forest pathologists involved (in addition to Kliejunas & Ko) that *P. cinnamomi* was the cause of the wide-spread 'Ōhi'a decline, the above cited study by Hwang & Ko focused once more on a carefully done quantitative analysis in adjacent healthy and decline areas at four different locations in the general dieback territory. Root samples were taken from 20 healthy 'Ōhi'a trees in each of the four healthy stands, and from 20 declining trees from each of the four declining stands. The results show that the amount of rootlets with *P. cinnamomi* was about the same for healthy and declining trees. No larger roots were infected, and quote "If *P. cinnamomi* was a major cause of 'Ōhi'a decline, one would expect it to be consistently associated with declining trees (first rule of Koch's Postulates)." The authors reemphasized the "nutrient deficiency hypothesis" stated in the 1974 paper above. They added that nutrient deficiency for pioneer 'Ōhi'a trees would increase with time because of the increase of stand density and tree size. (That seemed reasonable, but did not explain why there are so many co-existing healthy mature forest stands.)

One year later, an important conclusionary paper on the biotic disease hypothesis was published in 1979 as follows:

Papp, R. P., Kliejunas, J. T., Smith, R. J. & Scharpf, R. F. (1979). Association of *Plagithmysus bilineatus* and *Phytophthora cinnamomi* with the decline of 'Ōhi'a lehua forests on the island of Hawaii. *Forest Science* 25: 187–196.

Results: the endemic cerambycid beetle (*P. bilineatus*) and the root fungus (*P. cinnamomi*) were found to be associated with the decline of 'Ōhi'a trees. But neither appeared to trigger epidemic 'Ōhi'a decline. *"The two organisms which we studied acted independently, but attack by the cerambycid beetle was apparently encouraged by rootlet mortality or crown loss brought on by an* ***unknown stress factor****."*

Back in 1973, a well-known forest pathologist from New Zealand, F. J. Newhook, had been invited by the Institute of Pacific Islands Forestry to review the epidemic disease hypothesis. Dr. Newhook was aware of the findings of Kliejunas and Ko and restated a more refined version.

A different approach

In the 1974 proposal to the National Park Service Science Center in Mississippi (cited below) I stated Newhook's more refined disease hypothesis together with my alternative working hypothesis. Newhook's more refined disease hypothesis stated that " 'Ōhi'a dieback may be caused by root-pruning through a fungus and subsequent nutritional deficiency," still emphasizing a biotic disease concept. My succession hypothesis proposed that " 'Ōhi'a dieback is a normal phenomenon, a developmental stage in primary succession of an isolated island rain forest ecosystem." This rather bold succession hypothesis does not exclude Newhook's more refined pathological hypothesis. However, the succession hypothesis puts the 'Ōhi'a dieback in a broader context of forest dynamics.

Mueller-Dombois, D. (1974). *The 'Ōhi'a Dieback Problem: A Proposal for Integrated Research.* Honolulu: University of Hawai'i, Cooperative National Park Resources Studies Unit, Tech. Report 3. 20 p.

Copies of this report can be found in Hamilton Library, University of Hawai'i at Mānoa, Honolulu and at the Hawaiian Collection at the University of Hawai'i at Hilo.

With this proposal, I received funding from the National Park Service Science Center. I therafter began ecological research with three graduate students in January 1976: Jim Jacobi, Ranjit Cooray and Nadarajah

Balakrishnan. Together we established 42 plots (each 20 x 20 m) throughout the dieback territory in dieback and non-dieback stands. In each plot we analyzed the vegetation and soil, and measured all trees. The outcome was a vegetation map, a habitat classification, a tree health status classification, tree structural analyses with 26 plots set aside to be permanent, in which all trees were measured and labeled. We recognized five dieback types, and suggested a rainforest life cycle with five dynamic phases. The study was completed with a synthesis presented in a 1977 ʻŌhiʻa Decline Seminar together with the forest disease and insect pest researchers. It clarified that the problem was more complex than originally proposed by the epidemic disease hypothesis.

These results were initially reported in our third ʻŌhiʻa rainforest progress report together with the joint ʻŌhiʻa symposium (1977). Thereafter, they were republished in the following 1980 CTAHR booklet:

Mueller-Dombois, D., Jacobi, J. D., Cooray, R. G. & Balakrishnan, N. (1980). *ʻŌhiʻa Rain Forest Study: Ecological Investigations of the ʻŌhiʻa Dieback Problem in Hawaiʻi.* Honolulu: University of Hawaiʻi, College of Tropical Agriculture and Human Resources (CTHAR), Hawaiʻi Agricultural Experiment Station, Miscellaneous Publication 183. 64 p. [Republished from the final 1977 project report]

A restatement of the succession hypothesis into a more immediate working hypothesis for the study of ʻŌhiʻa canopy dieback appeared as chapter 7 in the following book:

Mueller-Dombois, D. (1980). The ʻŌhiʻa dieback phenomenon in the Hawaiian rainforest. In *The Recovery Process in Damaged Ecosystems,* ed. by J. Cairns, Jr. Ann Arbor: Ann Arbor Science Book. pp. 153–161.

Joint ʻŌhiʻa decline seminar 1977

The new Director of the Institute of Island Forestry, Charles S. Hodges, US Forest Service, organized a joint seminar of both research teams, inviting the forest disease experts (W. H. Ko, R. P. Papp, J.T Kliejunas and US Forest Service soil scientist H.B. Wood and botanist B. R. McConnel) to present their studies together with the UHM vegetation ecology team (Jim Jacobi, Ranjit Cooray, Grant Gerrish, and myself) in Honolulu on October 25, 1977.

The outcome was a shift from a disease to a forest ecological approach by the US Forest Service. Two new forest researchers were hired, the forest ecologist Ken T. Adee and the forest hydrologist Robert D. Doty. Subsequently, each of them produced excellent research reports that unfortunately were never published. The reports included:

Adee K. & Wood H. (1981). Regeneration and succession following canopy dieback in an 'Ōhi'a *(Metrosideros polymorpha)* rainforest on the island of Hawai'i. Paper presented at the Joint 'Ōhi'a Decline Seminar, Honolulu, October 25, 1977.

A 79-page document including 25 pages of maps, diagrams, and data. The authors supported almost completely the results of the 1980 CTAHR Misc. Publication 183 cited above.

Robert D. Doty's draft report entitled "Groundwater conditions in the 'Ōhi'a rainforest near Hilo" also remained unpublished. In contrast to expectation by most foresters, Doty's report shows that no accelerated or irregular flow of surface water could be detected in connection with the 'Ōhi'a forest decline. In other words, the watershed function was not impaired by the 'Ōhi'a dieback.

Following the shift of emphasis from a disease approach to an ecological approach, the US Forest Service came out with a final review, cited below:

Hodges, C. S., Adee, K. T., Stein, J. D., Wood, H. B. & Doty, R. D. (1986). *Decline of 'Ōhi'a* (Metrosideros polymorpha) *in Hawaii: A review.* Berkeley: US Department of Agriculture, Forest Service, Pacific Southwest Forest and Range Experiment Station. General Technical Report PSW-86. 22 p.

An excellent summary with color photos and diagrams, including a large-scale map as an appendix. It recognizes the same five dieback types identified earlier by the UHM team but with slightly different names, and adds two additional dieback types, 'Ōhi'a-Koa dieback and pubescent 'Ōhi'a dieback, the latter recognized as a first-generation dryland dieback. Towards the end the review recognizes the 'Ōhi'a decline is a typical decline disease (*sensu* Manion's death spiral analogy), and concludes by stating "In our view such broad definition of senescence in relation to declines [implying my view] in general helps little to identify the real precipitating causes of a particular decline problem and the term "synchronized cohort senescence" does not appear to describe in a meaningful way the etiology of 'Ōhi'a decline."

It is true that senescence does not identify the trigger causes of canopy dieback. Senescence is, however, considered as a major predisposing cause for dieback in the cohort senescence theory. It is a stage involving loss of vigor in the lifecycle of a population that can be brought on through aging, and when in combination with habitat constrains, it can be considered as "premature senescence." The USFS review also did not identify the "real precipitating causes." Thus, the critique is somewhat vague. By determining ʻŌhiʻa dieback as a typical decline disease, the US forestry team maintained that the phenomenon was an illness or abnormality. Moreover, it ignored the earlier finding by the disease expert team of the USFS (Papp, Kliejunas, Smith, Jr., and Scharpf) who stated that "the cerambycid [referring the endemic ʻŌhiʻa borer] was encouraged by rootlet mortality or crown loss brought on by unknown stress factors."

Unfortunately, this final USFS review also ended further research involvement of the Pacific Islands Forest Research Institute in the dieback/decline research. In the words of its new Head of the Institute, Dr. Charles Hodges, himself a mycologist and plant pathologist "the dieback problem can now be left to the ecologists" (*Honolulu-Star Bulletin*, January 30, 1984). Thus, work of myself and others continued, and the USFS team was dissolved.

Almost simultaneously with the USFS final report two explanatory papers were published:

Mueller-Dombois, D. (1986). Perspectives for an etiology of stand-level dieback. *Annual Review of Ecology and Systematics* 17: 221–243.

Mueller-Dombois, D. (1987). Natural dieback in forests. *BioScience* 37(8): 575–583.

Both papers are highlighted in Chapters 5 and 6 here in our book on the ʻŌhiʻa Rainforest.

A New Proposal and Research Plan 1980

Based on these findings that the widespread forest collapse was not caused by a new biotic disease but instead by an accumulation of natural stress factors, I wrote a new proposal to the NSF (National Science Foundation) in 1979. In this proposal, I argued that conditions for dieback can arise, when stands become stressed by substrate limitations (including nutrient limitations) and advanced age. Such stands then become predisposed to die. They only need a trigger to die. A trigger may simply be a tropical storm that strips off most of the foliage, which then cannot be replaced due to the chronic stress under which the predisposed stand has been suffering. Thus the elusive trigger, which we had not yet found could be a climatic instability.

NSF officers came for a visit in January 1980, and together we flew in a low flying aircraft across the dieback territory from North (near Laupahoehoe) to South (Kīlauea crater area) crossing a 20 x 10 mile rectangular area to view the mosaic patches of dying and live native 'Ōhi'a rainforest. The NSF team members expressed surprise about the Hawaiian rainforest. They were familiar with the tropical forest of Puerto Rico and had never seen such a simple tropical forest with a monospecific canopy, much of it in peril.

I received new funding for the research hypothesis that the dieback was a repeating event in the primary succession of an isolated rainforest. Subsequently, the NSF proposal was made generally available in form of a technical report below:

Mueller-Dombois, D. (1981). Spatial Variation and Succession in Tropical Island Rain Forests: A Progress Report. Honolulu: University of Hawai'i at Mānoa, Hawai'i Botanical Science Paper 41. 93 p.

This report states a new research plan of five integrated studies: analyses of spatial patterns of *Metrosideros polymorpha* dieback, field experiments (such as canopy removal, thinning of trees and fertilizer trials in a sample of stands that appear predisposed to die), study of nutrient dynamics, eco-physiological studies of 'Ōhi'a, and modeling of ecosystem processes.

Peter Vitousek became involved in the area of nutrient dynamics, and was later working on this topic with his team of graduate students and colleagues.

A principal finding of his team was nitrogen limitation early in primary succession of the Hawaiian rainforest and phosphorus limitation in late primary succession. This is well clarified, among many other findings, in his book:

Vitousek, P. (2004). *Nutrient Cycling and Limitation: Hawai'i as a Model System.* Oxford and Princeton: Princeton University Press, Princeton Environmental Institute Series. 223 p.

Other significant contributions emanating from the 1980 NSF grant include:

Jacobi, J. D. (1983). Metrosideros dieback in Hawai'i: A comparison of adjacent dieback and non-dieback rain forest stands. *New Zealand Journal of Ecology* 6: 79–97.

Lava flow structure was the primary difference between the two stands, the dieback growing on pāhoehoe, the non-dieback on 'a'ā lava. The latter also showed better drainage. "The results of this study lend support to the successional hypothesis of *Metrosideros* dieback" (from summary).

Burton, P. J. & Mueller-Dombois, D. (1984). Response of *Metrosideros Polymorpha* Seedlings to Experimental Canopy Opening. *Ecology* 65(3): 775–791).

In the ʻŌhiʻa displacement dieback area of Olaʻa Tract, twenty 100 m^2 plots were treated by defronding of dense tree fern sub-canopy. Shade-born seedlings responded slowly, many died in exposure to sunlight, but many new light-born seedlings came up in plots with drastic canopy removal and they grew rapidly into the sapling stage.

Gerrish, G. & Bridges, K. W. (1984). A Thinning and Fertilizing Experiment in *Metrosideros* Dieback Stands in Hawaii. Honolulu: University of Hawaiʻi at Mānoa, Hawaiʻi Botanical Science Paper 43. 107 pp.

The report describes three sites where stands appeared predisposed to dieback (at Saddle Road 18 mile marker, Puu Makaala, and Thurston Lava Tube). The experiment used three treatments, stand thinning, fertilizing, combined treatment and control. "Applied treatments of thinning and fertilizing did not reduce the rate of mortality of low vigor trees. At Saddle Road, only those trees treated with the combined thinning and fertilizing treatment showed a significant growth response" (p. 21).

Gerrish, G., Mueller-Dombois, D. & Bridges, K. W. (1988). Nutrient limitations and *Metrosideros* forest dieback in Hawaii. *Ecology* 69(3): 723–727.

The paper describes experiments to test Dr. Wen Ko's hypothesis that nutrient deficiency is the principal cause of tree death in stand-level dieback. The conclusion is that nutrient limitation is not the principal cause but that it can be a chronic stress in the predisposition to dieback.

The new NSF grant allowed for helicopter trips into deteriorating rainforest near 1000 m above Hāmākua Coast (July and August 1980, see Photos 4.25 and 4.26 and following story in Chapter 4). Here we were now enabled to study bog-formation and gap-formation dieback up close. In an area where trees appeared to have died a longer time ago and where larger treeless patches had been discovered from air photos, we set down, and made camp. We made two trips–July and August 1980–and spent several days doing plot work, which like elsewhere involved tree stand measurement, vegetation, and soil analyses as reported in the 1980 CTAHR booklet.

Monitoring the Hawaiian rainforest dieback

The following three published documents give the monitoring results:

Jacobi, J. D., Gerrish, G. & Mueller-Dombois, D. (1983). 'Ōhi'a dieback in Hawai'i: Vegetation changes in permanent plots. *Pacific Science* 37(4): 327–337.

This paper reports that in 1976 we established and analyzed throughout the dieback territory originally 42 plots, each 400 square meters in size. With NSF funding in 1979 20 more plots were added. Half of these were located in dieback and the other half in adjacent healthy forest stands. From these, 26 plots were selected as permanent plots. These were re-sampled in 1982 and their essential outcome is provided in the summary of this paper (see below).

See map (Chapter 6, p. 144), the dark circles indicate the permanent plots. This map also shows the areas of dieback types, the Wetland dieback, the Bog-formation dieback, and the 'Ōhi'a displacement dieback. The circles outside these dieback types refer to forest stands categorized as belonging to either Dryland dieback mostly south of the Saddle Road and Gap-formation dieback mostly north of the Saddle Road.

Quote from summary of our first plot re-monitoring study: "Nearly all the plots located in areas that originally experienced a drastic reduction of the tree canopy were found to have a large number of 'Ōhi'a seedlings and saplings."

A second re-monitoring was published after the International Botanical Congress in Berlin in 1987:

Jacobi, J. D., Gerrish, G., Mueller-Dombois, D. & Whiteaker, L. (1988). Stand-level dieback and Metrosideros regeneration in the montane rain forest of Hawaii. *GeoJournal* 17(2): 193–200.

From Abstract: "Seedling and sapling regeneration has been extremely vigorous in most of the sites that experienced a breakdown of the canopy while in stands with an intact, dense tree canopy no such regeneration occurred."

Boehmer, H. J. (2005). Dynamik und Invasibilitaet des montanen Regenwaldes auf der Insel Hawaii [Dynamics and Invasibility of Hawaii's Montane Rainforest]. Habilitation Thesis, Department of Ecology and Ecosystem Management, Technical University of Munich, Germany. 232 p. Includes six appendices.

Summary translated:

"We investigated the vegetation dynamics in permanent plots across the east flank of Mauna Loa and Mauna Kea. The focus was on the regeneration of the key species 'Ōhi'a lehua (*Metrosideros polymorpha*), an endemic tree that dominates the native forest canopy. This east-slope territory exhibited a spectacular ʻŌhiʻa dieback/decline during the 1970s into the mid-1980s. About 50 thousand hectares were affected by this native forest dieback/decline that was recognized as canopy dieback.

The regeneration model of Mueller-Dombois (1987), which predicts a direct response in form of rebirth of the same canopy species, was validated through our permanent plot study. The dead ʻŌhiʻa tree cohorts are now mostly replaced through young ʻŌhiʻa tree cohorts in the former dieback plots. They originated from waves of ʻŌhiʻa regeneration that were able to grow up after breakdown of the previous canopy. More than 50% of the plots exhibit a profoundly more vigorous ʻŌhiʻa canopy now after 25 years. However, some plots did not follow the predicted trend. These were subject to invasive plants in several ways. The dieback enhanced the invasiveness of some alien species in the mid 1970s. A problem was the alien strawberry guava tree (*Psidium cattleianum*), which displaced the predicted new ʻŌhiʻa cohorts in some places. Even the non-dieback forest was subject to invasion. The alien Kāhili ginger (*Hedychium gardnerianum*) formed a dense undergrowth layer under some ʻŌhiʻa canopies. This plant can prevent any reestablishment after canopy dieback. An earlier thesis that invasive plants would enter only disturbed ʻŌhiʻa forests is thus not applicable any more. Our case study on both the strawberry guava and Kāhili ginger has shown that these alien species can completely displace ʻŌhiʻa from its original site. This can lead to a complete change-over and biodiversity loss in the future, unless management attempts to control these aggressive aliens.

The trigger for the dieback was tentatively identified from rainfall records taken at two stations in the rainforest territory. Since the mid-1960s, an increased frequency of extreme rainfall events was noted in connection with ENSO anomalies. These, however, cannot cause a large area dieback as singular climatic disturbances.

The mechanisms identified during the ʻŌhiʻa dieback study may be applicable to Pacific forests in general, where key forest trees may be temporarily weakened by climate anomalies."

Boehmer, H. J., Wagner H. H., Jacobi, J. D., Gerrish, G. C. & Mueller-Dombois, D. (in press, 2013). Rebuilding after collapse: Evidence for long-term cohort dynamics in the native Hawaiian rain forest. *Journal of Vegetation Science.*

This paper reemphasizes the translated story above with full statistical support of the ʻŌhiʻa regeneration wave following dieback and renewal of the forest as discussed in Chapter 6 in this book.

Pacific forests with similar dieback

1981 was my sabbatical year involving research travel through the Melanesian Islands to gather material for a book with Dr. F. Raymond Fosberg on *The Vegetation of the Tropical Pacific Islands* (published by Springer-Verlag in 1998).

Wherever we (my wife Annette and I) traveled, I was particularly interested to discover if such stand-level dieback phenomena could be found as seen in Hawai'i. The first such encounter was a *Nothofagus balansae* dieback in New Caledonia that had not received any research attention. But it had the same symptoms as the *Metrosideros* dieback in Hawai'i, namely stand-level dieback restricted to the monospecific canopy. Similar diebacks were encountered in *Nothofagus pullei* stands on Mt. Giluwe and in *Araucaria hunsteinii* stands near Bulolo in Papua New Guinea. Both of these were later reported in one or the other of the three international forest dieback symposia, which I was able to organize with the help of supportive colleagues.

New Zealand: Another eye opener

Seminar at Victoria University of Wellington, December 1981. Bob Brockie got up after my talk in a Christmas seminar at Victoria University of Wellington saying, "you must have the Australian Possum or some other invading herbivore in the Hawaiian rainforest." Next day Bob Brockie took me to the *Metrosideros robusta* dieback in the research forest at Lower Hut. Here an observational study (by Entomologist Mead sitting many nights in a neighbouring tree) documented *Metrosideros robusta* dieback as caused by the introduced Australian bush-tailed possum (*Trichosorus vulpecula*).

Franz Joseph Glacier in Westland National Park. Ranger Gary McSweeney gave me an unpublished report by Veblen and Stewart on natural causes of *Metrosideros/Weinmannia* forest dieback. I took a photo of a slope with diagonal line that was said to be the stand of the glacier in 1750 (see Chapter 8, Photo 8.2, p. 186 in this book).

I had asked the question of fencing out possum from healthy lower forest separated from upper dieback forest along this diagonal line. I got a smile from Garry McSweeney. He went upstairs into the park library and got an unpublished report by Veblen and Stewart and another written by an entomologist blaming native insects for the dieback. Like in Hawai'i the disease explanation, so in New Zealand the possum explanation, is the politically correct explanation for stand-level dieback, still maintained today, but somewhat shaken by our 1983 dieback symposium in Dunedin (see below).

Craigieburn National Forest *Nothofagus solandri* var. *cliffortioides* (mountain beech) dieback: several photos and contact with Udo Benecke (see Chapter 8, Photos 8.5 through 8.7, pp. 189–191 in this book). Udo Benecke and I planned for an international dieback symposium; carried through as cited here:

The First International Dieback Symposium in New Zealand at the 15th Pacific Science Congress, 1983 (a follow-up of my New Zealand travel in 1982).

The Symposium was organized by Ross McQueen from Victoria University at Wellington and myself, and its proceedings published as follows:

Mueller-Dombois, D. & McQueen R. (eds.) (1983). Canopy Dieback and Dynamic Processes in Pacific Forests. *Pacific Science* 37(4).

Researchers contributed from Papua New Guinea (Frans Arentz, with *Nothofagus pullei* dieback on Mt. Giluwe, pp. 453–458) three from Australia (C. Palzer from Tasmania on "regrowth" dieback, pp. 465–470, G. A. Kile from southern Australia on *Armillaria* root rot in eucalypts, pp. 459–464 , and J. Walker, C. H. Thompson, and W. Jehne from Canberra and Brisbane on a very interesting progressive and retrogressive primary succession model similarly developed in Hawai'i, pp. 471–481). Eight papers were contributed from New Zealand. Several reported on abiotic natural dieback similar to that in Hawai'i (episodic dieback in the Kaimai Ranges by Jane & Greene, pp. 385–389; dieback in New Zealand's *Nothofagus* forests by John A. Wardle and R. B. Allen, pp. 397–404; forest instability and dieback of two-canopy dominant forests involving southern Rata and Kamahi (*Metrosideros umbellata & Weinmannia racemosa*) by Glenn H. Stewart & Thomas T. Veblen, pp. 427–431. Here instability related to occasional landslides that are regenerating with the same species in form of cohorts.

A strong defender of a biotic dieback cause in the same forest type was C. L. Batcheler, pp. 415–426, who was convinced of the overriding role of the introduced Australian possum *(Trichosurus vulpecula)* as foliage browser and tree killer. W. B. Shaw, pp. 405–414 emphasized the role of tropical cyclones causing patterns and structure in indigenous forests on New Zealand's North Island. He coined the term "ill thrift" for trees that linger without recovery after being hit, short of being uprooted (thereby supporting Sinclair's second dieback concept, see below).

Among the papers dealing with the Hawaiian dieback are the ones by Jacobi et al., pp. 327–337, giving the first results of 'Ōhi'a recovery from 26 permanent plots six years after the first assessment in 1976; Bill Evenson on the first climate analysis over the 'Ōhi'a dieback territory, pp. 375–384; Lani Stemmermann on the 'Ōhi'a species indicating changes from pubescent to glabrous leaf varieties along the primary successional gradient; Nadarajah Balakrishnan and Mueller-Dombois on soil nutrients in relation to habitat and dieback types, pp. 339–359; and Mueller-Dombois presented the first rendering of the cohort senescence theory as a four-level causal chain, pp. 321–322.

Receiving national & international attention

A literature review of five North American decline diseases was done in a graduate seminar by students of vegetation ecology in 1982. The diseases included the little leaf disease of shortleaf pine, western white pine pole blight, birch dieback, maple and oak declines. These were more or less unresolved mysteries, called "decline diseases." The outcome was published as follows:

Mueller-Dombois, D., Canfield, J. E., Holt, R. A., & Buelow, G. P. (1983). Tree-group death in North American and Hawaiian forests: A pathological problem or a new problem for vegetation ecology? *Phytocoenologia* 11(1):117–137.

This paper drew the attention of forest pathologists. In particular, the textbook author Paul Manion wrote a letter to me emphasizing that pathologists do not neglect ecology [of course not]. But he later graciously invited me to participate in his symposium of forest decline concepts at the Phytopathological Society. Prior to that we had met in Berlin at the International Botanical Congress as two keynote speakers on the "hot topic" at that time of *Waldsterben*, also later referred to as the "New-Type Forest Damage" in 1987.

Offering "cohort senescence" as an alternative to the "decline disease" concept, resulted in an invitation by Paul Manion (well-known Forest Pathologist) to present a paper in his symposium at the annual meeting of American Phytopathological Society in Grand Rapids, Michigan.

National Symposium on Forest Decline Concepts, St. Paul, Minnesota, August 7, 1990

From that symposium, a new book was published with the title *Forest Decline Concepts*, edited by Manion and Lachance (1992). It includes my contribution:

Mueller-Dombois, D. (1992). A natural dieback theory, cohort senescence as an alternative to the decline disease theory. In *Forest Decline Concepts*, ed by P. D. Manion and D. Lachance. St. Paul, MN: APS Press. pp. 26–37.

The Minnesota symposium clarified also that it was W. S. Sinclair (1965, 1967), who first developed the three-phase scheme to explain tree decline. He proposed: (1) Predisposing factors, followed by (2) Inciting factors and these by (3) Contributing factors, a sequence that has been adopted in Manion's death spiral model to explain decline disease. The three-phase causal sequence was proposed as a unifying theory also for cohort senescence (Mueller-Dombois 1988). The difference in the cohort senescence theory is an emphasis on two points, the aging factor as genetically programmed in each species, senescence, and on the outcome, implying recovery in form of a secondary succession. Habitat constraints such as nutrient imbalances or shifts in soil water regime can lead to premature senescence. This also has been recognized by Sinclair (1988) in a later paper cited below:

Sinclair, W. A. & Hudler, G. W. (1988). Tree declines: Four concepts of causality. *Journal of Arboriculture* 14(2): 29–35.

In this paper Sinclair and Hudler suggest four schemes for decline as applicable to different situations: (1) decline as a slowly progressing syndrome caused by a single factor, (2) decline as caused by a single shock, such as defoliation, that can be followed by "ill-thrift," whereby the tree lingers along without ever recovering its former vitality, (3) decline as explained in the three-phase process whereby the factors may be interchangeable, and (4) decline as applicable to trees growing in groups (cohorts) which develop and age together in form of synchronized senescence. Sinclair and Hudler write that this fourth scheme was advanced in recent years by vegetation ecologists (a reference to our Phytocoenologia paper cited above). He further states "Mueller-Dombois et al. [referring to the Phytocoenologia paper] adopted a narrow concept of disease in their discussions of synchronous cohort senescence, but neither this shortcoming nor the dissenting opinions of forest pathologists about the cause of ʻŌhiʻa decline [referring to the 1986 USFS final review] diminishes the intellectual attractiveness of the central idea in the cohort senescence concept."

XIV International Botanical Congress, Berlin: "Forests of the World"

The Proceedings were edited by Werner Greuter and Brigitte Zimmer and published by Koeltz Scientific Books, Koenigstein, Germany. Three keynote addresses on forest decline are published in the Proceedings volume:

Manion, P. D. (1988). Pollution and forest ecosystems. In *Proceedings 14th International Botanical Congress*, ed. by W. Greuter & B. Zimmer. Koenigstein, Germany: Koeltz Scientific Books. pp. 405–421.

Ziegler, H. (1988). Deterioation of forests in Central Europe. In *Proceedings 14th International Botanical Congress*, ed. by W. Greuter & B. Zimmer. Koenigstein, Germany: Koeltz Scientific Books. pp. 423–444.

Mueller-Dombois, D. (1988). Canopy dieback and ecosystem processes in the Pacific area. In *Proceedings 14th International Botanical Congress*, ed. by W. Greuter & B. Zimmer. Koenigstein, Germany: Koeltz Scientific Books. pp. 445–465.

The Second International Dieback/Decline Symposium

This symposium was organized by Uwe Arndt and myself within the frame of the IBC in Berlin, 1987. Its theme was "Stand-level Dieback and Ecosystem Processes—A Global Perspective" and was published in *GeoJournal* 17(2): 160-308, with German and North American forest decline research contributions on Waldsterben. The underlying idea was to compare forest dieback with industrial pollution called Waldsterben (forest death) in Germany and forest dieback without industrial pollution in Pacific countries. The symposium was conducted in two parts, (A) dieback/decline in forests with air pollution, (B) dieback/decline in forests without.

Dieback/decline in forests with air pollution

Contributions from Canada focused primarily on point source air pollution: Linzon (pp. 179–182), on sulfur emmisions of lead/zink smelters in Columbia River Valley at Trail, British Columbia (Kincaid and Nash, pp. 189–192), others on *Waldsterben* in Vermont and West Germany (Hoshizaki et al., pp.173–178). Twelve papers dealt with air pollution.

The conceptual model of Bernhard Ulrich, proponent of acid rain as an abiotic killer disease was extracted from his paper (with D. Murach, pp. 253–259) and placed on the cover of that issue of *GeoJournal.* It shows the essence of air pollution as caused by influx of gaseous pollutants. These are primarily sulphur dioxide (SO_2) and nitrous oxide (NO_2). These change into dry deposits damaging canopy foliage and then penetrate with dust during rainy weather into the soil, displacing the cations calcium, magnesium, and potassium with hydrogen protons in form of a mass exchange together with accelerated weathering of silicates. Other papers report on fumigation experiments with sulphur dioxide, nitrous oxide and ozone (mostly automobile exhaust pollutant) done in open-top chambers with rain shelters and soil lysimeters (Seufert & Arndt, pp. 261–270; Keller & Haesler pp. 277–278).

Dieback/decline in forests free of industrial air pollution

Almost the same number of papers (11) were presented from researchers that had worked on forest dieback/decline where industrial pollution was absent. They included two papers from New Zealand (Stewart and Rose, pp. 217–223; Ogden, pp. 225–230), one from Papua New Guinea (Arentz, pp. 209–215), one from Sri Lanka (Werner, pp. 245–248), two from Australia (Landsberg and Wylie, pp. 231–235; Davison, pp. 239–244), one from Japan (Kohyama, pp. 201–208). Four contributions were from Hawaiʻi (Mueller-Dombois, pp. 162–164, 249–251; Jacobi et al., pp. 193–200; Gerrish, pp. 295–299), and a final synthesis was given by Charles H. Gimingham from Aberdeen, Scotland (pp. 301–302).

In his synthesis, Charles Gimingham, a well-known ecologist from Scotland, concluded that "natural" dieback is an inherent feature in some forest vegetation, but not always for obvious reasons. He refers to an earlier paper by Jones who observed that most temperate forests are composed of tree cohorts established over areas of varying size where the previous stand had suffered dieback (see Jones (1945)). The structure and reproduction of virgin forests of the North Temperate Zone. *New Phytologist* 44: 130–148). Gimingham lists several other literature examples and says that forest dieback/decline is "not to be explained simply as the effect of this or that pollutant. Assessment must be made of the extent to which death from "natural causes" is to be expected, on the basis of normal population dynamics."

Gimingham concludes by stating "The problem is perhaps the most complex ever to have confronted applied ecologists, and if the forest resources are to be maintained, it is one which demands urgent attention."

Two years later another European Dieback Symposium dealt with soil nutritional stresses as predisposing factors.

International Union of Forest Research Organization (IUFRO) Symposium, Freiburg, Germany, 1989, organized by Heinz W. Zoettl & Reinhard F. Huettl

The symposium was structured in four parts: I. Environmental pollution and forest nutrition; II. Natural stresses and forest nutrition; III. Effects of liming and fertilization in the forest ecosystem; IV. Strategies for increasing health and productivity of forests.

I was invited to contribute a paper to part II of the symposium:

Mueller-Dombois, D. (1988). Forest decline and soil nutritional problems in Pacific areas. In *Management of Nutrition in Forests under Stress,* ed. by H. W. Zoettl & R. F. Huettl. Freiburg, Germany: Kluwer Academic Publishers. pp. 195–207.

This paper was based on my first-hand insight gained on the New England Dieback also named Rural Eucalypt Dieback in eastern Australia. I visited this area together with dieback researchers Jill Landsberg and Meg Lowman. Here, the indigenous eucalypt forest had been reduced to small fragments to make way for sheep pastures.

Jill Landsberg had participated in the Berlin symposium and there presented an elaborate conceptual model of the rural eucalypt dieback in Western Australia published in *GeoJournal* 17(2): 223.

Meg Lowman graciously presented me with her book here cited:

Heatwole, H. & Lowman, M. (1986). *Dieback—Death of an Australian Landscape.* French Forests, NSW: Reed Books Pty. Ltd. 150 pp.

This is a well-illustrated nine-chapter book, which answers three questions (reminiscent of Robert Nelson's questions in promoting the disease hypothesis in Hawai'i): What is dieback? What are its causes? What can be done about it? The book adds thoughtful sections on: life and death of a tree, insects and dieback, and environmental change and dieback.

I could not have had a better introduction when in 1988 I visited Meg Lowman in Australia. I had informative discussions with her and Jill Landsberg in the field, and also with local foresters there, who held native insects responsible for the spectacular New England dieback.

Some people had advanced the idea that the native koala bear (*Phascolarctos cinereus*) was involved in killing remnant groups of eucalypts, which is true in this area. However, blaming this poor little marsupial tree climber for defoliating his drastically reduced resource is not unlike the officially more accepted version that native insects were the culprits. Yet, not only the extreme forest reduction to forest remnant groups was a predisposing cause for dieback, another was the broadcast application of phosphate fertilizers from low-flying aircraft. Fertilizer from planes had spread into the forest remnants making the eucalypt foliage more palatable and thereby increasing the native insect herbivory. This clearly was a case of ecosystems out-of-balance due to overextending one resource (pastures with introduced pasture grasses needing frequent fertilization) to the neglect of the other resource (native forest).

The Third International Dieback/Decline Symposium, Hilo, Hawaiʻi, June 2–6, 1991, as a satellite symposium of the XVII Pacific Science Congress in Honolulu

Most symposium presentations appeared in the 1993 Springer book:

Huettl, R. F. & Mueller-Dombois, D. (eds.) (1993). *Forest Decline in the Atlantic and Pacific Region*. Berlin/Heidelberg: Springer-Verlag. 366 p.

Atlantic and Pacific forest researchers came together to present results and discuss their studies of dieback/decline from their respective areas. Atlantic forest dieback papers were contributed from Canada (on sugar maple decline in Quebec by Cote et al., pp. 162–174); three from the US (Hertel et al., pp. 54–65; Johnson, Lovett et al., pp. 66–81, explaining a big Integrated Forest Study (IFS) on Atmospheric Pollution, Forest Nutrition, and Forest Decline under the Electric Power Research Institute (EPRI) coordinated by L. F. Pitelka with an IFS advisory panel chaired by the well-known US ecologist Bill Reiners, and involving numerous scientists from all over the US, including a few from Norway and Canada, a program reminiscent of the US/IBP; further by Adamowicz, Skelly and McCormick, pp. 144–161), one each from Switzerland (Haemerli and Schlaepfer, pp. 3–17), France (Landmann pp. 18–39), Great Britain (Innes, pp. 40–53), and Finland (Raitio, pp. 132–143). Seven papers were from dieback/decline studies done in Germany. They included Kreutzer, p. 82–96 on the role of nitrogen; Huettl, pp. 97–114 on Mg deficiency as a "New phenomenon;" Senser and Hoepker, pp. 115–131 on K deficiency and "Acute Yellowing;" Fink, pp. 175–188 showing colored micrographs with needle yellowing resulting from Mg deficiency and others with K deficiency. Electron micrographs demonstrate breakdown of phloem with Mg deficiency in needle tissue; Kottke et al, pp. 189–201 focusing on micorrhizal fungi; Kenk, pp. 202–215 on growth analyses in "declining" forests; Otto Kandler, pp. 216–226 on demonstrating with photos and data the recent episode of *Tannensterben* [fir decline] in Bavaria. In all 15 contributions came from areas where industrial pollution was the potential cause of forest decline. But interestingly, *Waldsterben* due to acid rain was toned down in all papers, and the focus was more on nutritional imbalances for various reasons, some as yet unexplainable.

Contributions from areas free of industrial pollution included one from Bhutan (Donaubauer, pp. 332–337), one from New Guinea (Enright, pp. 321–331 on *Araucaria* dieback), one from Australia (Lowman and Heatwole, pp. 307–320 on eucalypt dieback with three spectacular color photos), and four from New Zealand: Ogden et al., pp. 261–274 on episodic mortality in several tree taxa and forests in New Zealand's dynamic volcanic landscape, fully supporting the cohort senescence theory; Hosking, pp. 275–279 on *Nothofagus* decline in New Zealand from a pathologist's perspective; Simpson pp. 280-292 on the sudden decline of cabbage trees (*Cordyline australis*) in New Zealand; Hunter, pp. 293–306 on *Pinus radiata* deline in New Zealand plantations on New Zealand's North Island as caused by inadequate nutrition, primarily P & Mg deficiency. Hunter makes the point that needle yellowing in Radiata pine plantations has shown to be similar to the "New-Type Forest Damage" in the Atlantic forests. But in Radiata pine, Mg deficiency is the species' inability to absorb a sufficient supply from the soil, while other species, for example *Eucalyptus regnans* and *Acacia dealba* are more efficient Mg absorbers. Thus appropriate species/site matching for tree plantations is an important factor, certainly not a new discovery.

Four contributions were from Hawai'i (Jacobi, pp. 236–242 on *Metrosideros* dynamics; Gerrish, pp. 243–250, on tree senescence and premature senescence using a life-history-carbon-balance model; Mueller-Dombois, pp. 229–234, a brief summary of the *Metrosideros* canopy dieback in Hawai'i and pp. 338–348 providing for a global synthesis, and Jeltsch and Wissel, pp. 251–260 on modeling stand-level dieback).

At this time the notion of *Waldsterben* all over Europe was scaled down to *Neuartige Waldschaeden* (New-Type Forest Damage). The most outstanding symptom was needle yellowing in spruce and fir as an abiotic disease. This symptom had appeared over wide areas. It was identified as magnesium deficiency. Initially this was thought to be attributable to acid rain. But it turned out to be also in areas without acid rain. Mg deficiency was found also in *Pinus radiata* plantations on the North Island of New Zealand (Hunter, pp. 293–306). The cause could be the soil lacking enough Mg to sustain the health of foliage, but the simultaneous appearance in Europe and New Zealand made Reinhard Huettl, pp. 97–114 suggest climatic causes, such as global warming, episodes of drought, and CO_2 enhancement of the atmosphere.

Forest researchers who found the Hawaiian model broadly applicable to forests in New Zealand and New Guinea are cited below:

Ogden, J. (1988). Forest dynamics and stand-level dieback in New Zealand's *Nothofagus* Forests. *GeoJournal* 17(2): 225–230.

Here Ogden describes the regeneration systems of three NZ beech species, mountain beech *(Nothofagus solandri* var. *cliffordioides)*, red beech *(N. fusca*), and silver beech (*N. menziesii).* Mountain beech is relatively short-lived and light demanding. Its regeneration system depends on larger area canopy breakdown. Red beech is a taller and longer lived tree, which can maintain its population by gap-phase regeneration. Silver beech is more shade tolerant and has a still longer lifespan. When growing in mixture, silver beech tends to replace the other two beech species in the absence of canopy breakdown. He shows two dieback/regeneration cycles diagrammatically (p. 226 and p.228) implying cohort stands and synchronizing triggers, which set off canopy dieback in predisposed cohort stands.

Stewart, G. H. (1989). Ecological considerations of dieback in New Zealand's indigenous forests. *NZ Journ. of Forestry Science* 19 (2/3): 243–249.

"Three types of factors influence the dieback of forest stands—factors that predispose stands, trigger factors that initiate dieback, and factors that contribute to further decline. All known examples of dieback in New Zealand *Nothofagus* spp., *Metrosideros* spp. and beech/hardwood forests can be explained using this three-factor framework."

Ogden, J., Lusk, C. H. & Steel, M. G. (1993). Episodic mortality, forest decline and diversity in a dynamic landscape: Tongariro National Park, New Zealand. In *Forest Decline in the Atlantic and Pacific Region,* ed. by R. F. Huettl & D. Mueller-Dombois. Berlin/Heidelberg: Springer-Verlag. pp. 261–274.

The authors adopted the terms "cohort senescence" and "cohort structure." They also refer to "bog-formation dieback" as loss of site for tree growth, and to "replacement dieback" as simple self-replacement (implying the dieback species) and "displacement dieback" as replacement of dieback by other species.

"The cohort senescence phenomena exhibited by *Nothofagus solandri* in the Tongariro National Park are similar to those of *Metrosideros polymorpha* in Hawaii: all types of dieback occur, but replacement by the same species is the most common" (p. 272).

Enright, N. J. (1993). Group Death of *Araucaria hunsteinii* K. Schumm (Klikii pine) in a New Ginea rainforest. In *Forest Decline in the Atlantic and Pacific Regions*, ed. by R. F. Huettl & D. Mueller-Dombois. Berlin/Heidelberg: Springer-Verlag. pp. 321–331.

Size class distribution of klinkii pine, a major emergent tree in some PNG rainforests, shows a normal bell-shaped curve (see Fig. 1, p. 323) demonstrating a cohort stand. Enright discusses four hypotheses to explain stand-level dieback of klinkii canopy trees: 1. Cohort senescence, 2. Soil nutrient limitation, 3. Changing water relations, 4. Pathogens, disease, herbivory.

He concludes "there was no measurable stress in trees prior to rapid foliage loss and death" (p. 330).

The trigger he assumes to be a native termite (*Coptotermes elisae*). There was no landscape disturbance. Thus, the trees must have been predisposed to dying due to advanced aging or senescence.

Conclusions

A global review of forest dieback was published by FAO in the following document:

Ciesla, W. M. & Donaubauer, E. (1994). *Decline and Dieback of Trees and Forests: A Global Overview.* Rome: Food and Agriculture Organization of the United Nations, FAO Forestry Paper 120. 90 p.

The authors added to the Atlantic and Pacific forest dieback/declines, several others from Africa, China, India, and Latin America. They use the three-phase approach of predisposing, inciting, and contributing causes and test each dieback case for the applicability of the current dieback/decline theories. In the conclusion they state "For evaluating the impact of new anthropogenic stresses such as air pollution, climate change and biotic impoverishment on forests, it is important to understand the natural process of forest dynamics. Only then will it be possible to untangle the real impact of human influences on forest decline and dieback" (Mueller-Dombois 1992: 36). A natural dieback theory, cohort senescence, as an alternative to the decline disease theory.

Finally, I come back to an early classic on forest dieback:

Sprugel, D. G. (1976). Dynamic structure of wave regenerated *Abies balsamea* in the north-eastern United States. *Journal of Ecology* 64: 889–911.

The paper emphasizes the outcome of the dieback, the form of recovery; in this case it is auto-succession (replacement of the dieback population with a new population wave of the same species). The kind of recovery is in fact the critical difference between the forest decline disease and natural/normal dieback in form of senescence including premature senescence. Decline ends in a death spiral with uncertain outcomes, while dieback in form of canopy dieback ends in rejuvenation in form of auto-succession, unless new invasive species interfere.

Stripped from complexities, canopy dieback following cohort senescence is simply the turnover from one generation to the next. This applies to forests with only one or few dominant canopy trees whose progeny does not develop into trees until the canopy gives way to the light resource needed for the progeny to grow up. It is a feature related to biotic impoverishment associated with limitations in functional traits. Shade-intolerant canopy populations such as the Hawaiian ʻŌhiʻa (*Metrosideros polymorpha*) cannot renew themselves in form of auto-succession until the closed parental canopy collapses. The turnover process is similar to the gap-phase replacement in forests with more species sharing the canopy, except that patches created in forests with dieback dynamics are usually much larger as they depend primarily on the spatial extent of cohorts in such forests.

Canopy dieback logo from project reports, 1975 to 1985.

Dieback dynamics in Manitoba: A final note

Prior to my arrival in Hawaiʻi in 1963, I worked as a Research Officer in the Department of Forestry Canada from 1958–1963. My research task was to develop a forest-land classification for the Rainy River Section of the Great Lakes-St. Lawrence Forest Region (Rowe 1959) in southeast Manitoba, known as the Sandilands. A major forest cover type here is the monospecific Jack pine (*Pinus banksiana*) forest that is widely distributed over former beach deposits, glacial outwash, inland dunes and recessional moraines, i.e. remnant landforms of the former glacial lake Agassiz (see Figure 11.2, p. 322, in Mueller-Dombois & Ellenberg 1974, 2002).

Since my task was to investigate forest soil-vegetation relationships with measures for tree growth capacities, I focused on spatial distribution rather than dynamic relationships (Mueller-Dombois 1963, 1964a, 1964b, 1965). However, by studying many Jack pine and Black spruce (*Picea mariana*) stands, it became abundantly clear also that the distribution of forest vegetation types was not influenced only by soil-substrate variations, but also by the major disturbance regime in this continental environment, namely recurring fires.

Jack pine stands in particular exhibit dieback dynamics that are functionally similar to the dieback dynamics in the 'Ōhi'a rainforest. Jack pine stands form cohorts (generation stands) that typically originate after canopy fires in senescing stands. In aging Jack pine stands, dry branches accumulate on the forest floor together with serotenous (tightly closed) cones. Many cones also remain attached to the branches in the senescing pine crowns. They open when scorched by fire and then release their seeds, which thereafter germinate forming a new cohort forest. Stand-level senescence in Jack pine is a common predisposition to forest fires in southeast Manitoba. The fires act as the trigger factor for stand-level collapse followed by stand-level rebirth.

Thus, our *'Ōhi'a Lehua Rainforest* book subtitle (p. iii): "The Story of a Dynamic Ecosystem with Relevance to Forests Worldwide" has definite application. It may also provide advice for forest management in landscapes prone to forest fires, namely to plan for renewing its woody vegetation before stands enter the senescing life stage. If fire does not arrive on time, bark beetles may start the trigger effect, and subsequent fires become even worse.

References

Rowe, J. S. (1959). *Forest regions of Canada*. Canada Department of Northern Affairs & Natural Resources. Forestry Branch Bulletin 123.

Mueller-Dombois, D. (1963). *Techniques for studying soil-water growth relations on an artificial slope*. Dept. of Forestry Canada. Forest Research Branch Contrib. No. 561: 153–161.

———. (1964a). The forest habitat types in southeastern Manitoba and their application to forest management. *Canadian Journal of Botany* 42: 1417–1444.

———.(1964b). Effect of depth to water table on height growth of tree seedlings in a greenhouse. *Forest Science* 10(3): 306–316.

———. (1965). Eco-geographic criteria for mapping forest habitats in southeastern Manitoba. *The Forestry Chronicle* (Canada) Vol. 41 (2): 188–206.

Mueller-Dombois, D. & Ellenberg, H. (1974). *Aims and Methods of Vegetation Ecology*. John Wiley & Sons, NY. 547p. Republished 2002 by Blackburn Press, NJ.

Appendix D

A Local Perspective and Timeline of the ʻŌhiʻa Dieback/Decline

as reported from Newspaper Articles, Memoranda, and Symposia in Hawaiʻi

Dulce Rieza Belen
Dieter Mueller-Dombois
Hans Juergen Boehmer

August 1965: Photograph by Roger Baldwin, showing a *Metrosideros* dieback stand near Mile 15 along Saddle Road/Big Island/Hawaiʻi.

May 1970: "Many Island Trees Fall Victims to Imports," *Star-Bulletin & Advertiser* (May 31). R. E. Nelson on futility of saving native forests in the Hawaiian Islands (viewpoint 1).

June 1970: "Ecologist says Isle Forests can Survive," *Star-Bulletin & Advertiser* (June 14). Dieter Mueller-Dombois [DMD]'s response to Nelson's article (viewpoint 2).

July 1970: "Hawaii joins the search for survival," *Honolulu Star-Bulletin* (July 13). Announces NSF funding for IBP research in Hawaiʻi, including two photos of landscape-level tree mortality: upper montane Koa and montane Ohia Rainforest on the east side of Mauna Kea.

August 1970: Inspection tour with Dr. Ivan Buddenhagen, resulting in two viewpoints (hypotheses):

1) Alien (biotic) disease—Buddenhagen's hypothesis, supported by Petteys et al. (1975);

2) Natural phenomenon (abiotic stress)—DMD's hypothesis, based on Mueller-Dombois & Krajina (1968).

August 1973: Dr. Buddenhagen memo (August 30). Lists 10 points emphasizing hypothesis 1 with *Phytophthora cinnamomi* as the principal cause.

February 1974: "Ohia epidemic worsens," *Honolulu Advertiser* (February 7). "Hawaii's largest ohia forest—a 200,000 acre section of the Big Island—will be completely destroyed by 1985 if a solution is not found to halt a rapidly advancing blight. This was the prediction yesterday by State Forester Tom Tagawa to the Hawaii County Council on the State and Federal attempt to stop the epidemic."

March 1975: US Forest Service and NPS responds to DMD's 1974 proposal for integrated research (three points by plant pathologist Robert Scharpf and one point by NPS Chief Scientist, Western Region, O. L. Wallis saying that funding for research on the sucession hypothesis is unlikely).

May 1975: "Death of Ohia Forests a Frightening Prospect," *Honolulu Star-Bulletin* (May 17). A five-year retrospective on research focusing on hypothesis 1, the disease hypothesis.

June 1975: New proposal by Robert Scharpf (May 23) focuses on *Phytophthora cinnamomi* as the most likely cause, attached to NPS letters by O. L. Wallis to Robert Nelson, saying that DMD's proposal will be funded on a reduced level and with modifications (after DMD applied political pressure via Councilwoman Merle Lai).

August 1975: "Ohia dieback fight gets more funding," *Honolulu Star-Bulletin* (August 7). Article gives three hypotheses—after DMD receives $55,000 for 3 years from the NPS Science Center. The three hypotheses are:

(1) The dieback is caused by a biotic disease involving a killer fungus and /or an insect pest

(2) The dieback is caused by soil nutritional limitations

(3) The dieback is a normal dynamic phenomenon repeatedly occurring in long-term primary succession of an isolated rainforest ecosystem.

December 1975: "Hawaiian rain forests are in trouble," *Christian Science Monitor*, Boston. The problem is seen as becoming more complicated because several researchers gave different opinions.

March 1976: "Ohia blight caused by insect, fungus," *Honolulu Star-Bulletin* (March 13). Article attached to March 16 memo by Dr. Oliver Holtzmann to Drs. Ko and Kliejunas (and others, including DMD) with the request to make no comments to the press unless information is channeled through University of Hawai'i Pathology Department; Holtzmann says, if they were to make comments to press, there would be consequences (comment: Dr. Ko lost his research funds at this time).

March 1977: "Ohia trees are dying," *Honolulu Star-Bulletin* (March 7). Reports on S. C. Hwang's finding that *Phytophthora cinnamomi* is not a primary cause. Dr. Hodges, however, says it could be a contributing factor; Dr. Scharpf says this is the most complicated problem he has ever seen. The prediction that complete elimination of *Metrosideros* forest is possible in 15–25 years is repeated. DMD says that research has made progress by eliminating some possibilities. Robert Nelson pointed to the broken balance of nature. All scientists agree that further research is needed.

April 1978: "Ohia forest report," *Honolulu Star-Bulletin* (April 10). Reports on the first three years' results of work concerned primarily with hypotheses 2 saying that "Ohia rainforest is not an endangered ecosystem."

August 1979: "UH scientist receives ohia studies grant," *Honolulu Star-Bulletin.* (August 20). Reports on DMD's NSF award of $425,000 for a 3-year study.

April 1983: "Ohia dieback isn't a death knell," *Honolulu Star-Bulletin* (April 29). Reports on three years of NSF research with 3 graduate students and a conclusion made after DMD's research travel to Papua New Guinea, New Zealand, Australia and the 15th Pacific Science Congress Symposium "Canopy Dieback and Dynamic Processes in Pacific Forest," held in Dunedin, New Zealand, and published in *Pacific Science* 37(4).

January 1984: "Dieback," *Honolulu Star-Bulletin* (January 30). Charles Hodges says, the "dieback problem can now be left to ecologists."

September 1984: "Diagnosing forest illnesses," *Honolulu Star-Bulletin* Article (September 10). Reports on research connections made with North American and European forest decline research. This is the last local article.

1987: 2nd International Symposium (after New Zealand in 1983) on Forest Decline at the 14th International Congress of Botany in Berlin, Germany; proceedings published as "Stand Level Dieback and Ecosystem Processes in a Global Perspective" in *GeoJournal* 17 (2) (1988).

1991: 3rd International Symposium on Forest Decline, "Forest Decline in the Atlantic and Pacific Regions," based on Pacific Science Congress Symposium in Hilo (June 2–6). Proceedings published by Springer, Berlin/New York (ed. by Huettl & DMD 1993).

1992: "Forest Health Issues on a Global Perspective," convened by John M. Skelly and published in *Env. Toxicology and Chemistry* 11(8), including an article by DMD "A global perspective on forest decline." (pp. 1069–1076).

1994: *A Global Overview of Decline and Dieback of Trees and Forests* was published in FAO Forestry paper 120 (90 p.) by William M. Ciesla and Edwin Donaubauer. It evaluates case examples from Europe, North America, Africa, Asia, Australia, New Zealand, the Pacific Islands, Latin America, and the Caribbean, and concludes with a statement "For evaluating the impact of new anthropogenic stresses such as air pollution, climate change and biotic impoverishment on forests, it is important to understand the natural processes of forest dynamics. Only then will it be possible to untangle the real impact of human influences on forest decline and dieback" [from Mueller-Dombois, "A Natural Dieback Theory, Cohort Senescence, as an Alternative to the Decline Disease Theory," in *Forest Decline Concepts*, edited by Paul D. Manion and Denis Lachance (St. Paul, MN: American Phytopathological Society, APN Press), p. 36].

Since 2001: "Regeneration and invasibility of the Hawaiian *Metrosideros* rainforest," research fellowship Hans Juergen Boehmer, funding by the German Research Foundation (DFG); the primary objective of study is to document the regeneration of native ʻŌhiʻa lehua (*Metrosideros polymorpha*) forest after dieback and ecosystem-level consequences of biological invasions.

About the Authors

Dieter Mueller-Dombois (PhD 1960) drew scientific attention to the Hawaiian rainforest dieback in the early 1970s. While the US Forest Service had expected and then thoroughly researched the dieback as a new disease problem, Mueller-Dombois suggested an ecological study on the successional dynamics of this forest ecosystem. With support from several research grants he worked personally in the field with his graduate students for four decades throughout the Hawaiian rainforest. This included helicopter trips into remote areas and work across the island chain from the youngest to the oldest high island. He organized three international symposia on dieback dynamics and authored and co-authored numerous scientific papers, including several books, among them *Island Ecosystems: Biological Organization in Selected Hawaiian Communities* (1981), *Vegetation of the Tropical Pacific Islands* (1998), *Hawai'i, the Fires of Life* (2007), and *Biodiversity Assessment of Tropical Island Ecosystems: PABITRA Manual for Interactive Ecology and Management* (2008).

Photo by Olivier Koning

James D. Jacobi (PhD 1990) has been conducting research in Hawai'i for over forty years with primary emphasis on studying the ecology and status of native plant and bird species and communities, as well on the impacts of invasive species on Hawaiian plant communities. A major focus of his research has also been mapping the distribution of plant communities throughout the Islands. He was a graduate student under Professor Dieter Mueller-Dombois, and worked on both the International Biological Program (IBP) and the 'Ōhi'a dieback research projects. He received his PhD from the University of Hawai'i at Mānoa in 1990 and is now a biologist with the US Geological Survey's Pacific Island Ecosystems Research Center.

Photo by Jutta Pscherer

Hans Juergen Boehmer (PhD 1998) started his research in Hawai'i's rainforests in 1999 after he was accepted as a Post Doctoral Fellow at the Botany Department, University of Hawai'i at Mānoa, Honolulu. His research focuses on the long-term dynamics of montane 'Ōhi'a forests on the Big Island. He received several research grants for detailed studies on the regeneration of 'Ōhi'a after dieback, and the interactions of 'Ōhi'a population dynamics with invasive alien plant species. In 2005 he completed his habilitation thesis on "Dynamics and Invasibility of Hawaii's Montane Rainforests" at the Department of Ecology and Ecosystem Management, Technical University of Munich, Germany, where he was appointed Adjunct Associate Professor (PD) of Landscape Ecology thereafter. Currently he is working as a senior research scientist and managing director of the Interdisciplinary Latin America Center (ILZ) at the University of Bonn, Germany.

Jonathan P. Price (PhD 2002) has researched Hawaiian ecosystems for over fifteen years, and focuses primarily on the biogeography and landscape ecology of native Hawaiian plants. In 2002, he received his PhD in Geography from the University of California at Davis, followed by two years as a post-doctoral researcher at the Smithsonian Institution in Washington, DC and three years working at the US Geological Survey's Pacific Island Ecosystems Research Center. He is currently an associate professor of Geography and Environmental Studies at the University of Hawai'i at Hilo.

Book Reviews

The book is a remarkably good choice for acquisition by libraries where there is interest in ecology and natural history. The book has a place on the scientist's desk and it presents excellent ecological science, in the hands of the general visitor to Hawai'i. The book is a fascinating read on the Islands' natural history, and in student hands a textbook on Island Ecology.

László Orlóci, FRSC
Emeritus Professor of Statistical Ecology
University of Western Ontario, London, Canada

This story of a dynamic ecosystem exemplifies a unique paradigm switch from tree group dieback as seen from a pathological perspective to revealing it as a normal process in ecological succession. The process relates to the turnover of tree generations (cohort stands) forming sub-units in a forest, whereby canopy collapse provides for the release of the next generation. The book reviews five decades of impressive research in words and pictures with results basic to forest science. This fundamental study provides a baseline for forests currently experiencing global change.

Reinhard F. Huettl, Prof. Dr. Dr. h.c.
Wissenschaftlicher Vorstand
Helmholtz-Zentrum Potsdam
Deutsches GeoForschungsZentrum GFZ

Professor Mueller-Dombois and his collaborators have created an extraordinary book—in part a guide to and explanation of what you see in Hawai'i Volcanoes National Park, in part a fascinating story of a rich career spent in Hawaiian forests, and in part an accessible and engaging discussion of the science of ecology. I recommend it for travelers and scientists alike.

Peter Vitousek, PhD
Clifford G. Morrison Professor
of Population & Resource Studies
Stanford University

This dieback story of ʻŌhiʻa lehua forest highlights all the elements of good ecological science: (1) teamwork, (2) experience gained over time, (3) many peer-reviewed publications, (4) an infusion of funds for team members—especially students—who require financial support, (5) long-term ecological research with sustained enthusiasm, (6) the successive approximations of truth over time, meaning the necessity of project leaders to have patience and to retain an open mind. These six elements are common to all good modern ecological studies. Indeed, the scientific method itself could not be better summarized than by reading the chapters in this book .

The research peeled back paradigm after paradigm in the 1960s and 1970s, like so many layers of an onion. Those rejected paradigms have been replaced by new ones that have only become stronger over the past few decades. Professor Mueller-Dombois and his team were ahead of the curve on the nature of plant communities, succession, disturbance, and climax vegetation. Indeed, his work could not have been accomplished and accepted so readily without the parallel publication of *Aims and Methods of Vegetation Ecology* (1974). The field ecological approach taken for ʻŌhiʻa lehua research was based on that textbook. The book's reprinting in 2002 demonstrates the tremendous and unique value of that textbook, still able to support novel and important research despite the passing of nearly 40 years.

Michael Barbour, PhD
Emeritus Professor of Plant Ecology
John Muir Institute of the Environment
University of California at Davis

Finally, a book dedicated to the ʻŌhiʻa lehua rainforest! Often overshadowed by the regal Koa, the ʻŌhiʻa, as Dr. Sam Gon's elegant prologue expresses, is the key species of our native landscape and deserves the overview and attention found here. Especially pleasing is the clear and concise treatment of the complex and globally significant detective work into the ʻŌhiʻa canopy dieback process. Beautifully illustrated, heavily annotated and targeted for the general reader, this work is a must-have for anyone with an interest in Hawaiian natural history or a passion for our beloved Hawaiʻi Nei.

Rob Pacheco
Naturalist and Founder
Hawaii Forest & Trail

Made in the USA
San Bernardino, CA
14 May 2013